Local and Urban Governance

Series Editor

Carlos Nunes Silva, LISBOA, Portugal

This series contains research studies with policy relevance in the field of sub-national territorial governance, at the micro, local and regional levels, as well as on its connections with national and supranational tiers. The series is multidisciplinary and brings together innovative research from different areas within the Social Sciences and Humanities. The series is open for theoretical, methodological and empirical ground breaking contributions. Books included in this series explore the new modes of territorial governance, new perspectives and new research methodologies. The aim is to present advances in Governance Studies to scholars and researchers in universities and research organizations, and to policy makers worldwide. The series includes monographs, edited volumes and textbooks. Book proposals and final manuscripts are peer-reviewed.

The areas covered in the series include but are not limited to the following subjects:

- Local and regional government
- Urban and metropolitan governance
- Multi-level territorial governance
- Post-colonial local governance
- Municipal merger reforms
- Inter-municipal cooperation
- Decentralized cooperation
- Governance of spatial planning
- Strategic spatial planning
- Citizen participation in local policies
- Local governance, spatial justice and the right to the city
- Local public services
- Local economic development policies
- Entrepreneurialism and municipal public enterprises
- Local government finance
- Local government and sustainable development
- Anthropocene and green local governance
- Climate change and local governance
- Smart local governance

The series is intended for geographers, planners, political scientists, sociologists, lawyers, historians, urban anthropologists and economists. **Local and Urban Governance—now indexed in Scopus**

Abraham R. Matamanda • Tazviona R. Gambe •
Johannes I. Bhanye • Tafadzwa C. Maramura •
Oratilwe Annastacia Khoza

Urban Water Governance in Postcolonial Zimbabwe

A Resilience and Sustainability Perspective

Springer

Abraham R. Matamanda
Department of Urban & Regional Planning
University of the Free State
Bloemfontein, South Africa

Tazviona R. Gambe
Centre for Development Support
University of the Free State
Bloemfontein, South Africa

Johannes I. Bhanye
African Climate and Development Initiative
University of Cape Town
Cape Town, South Africa

Tafadzwa C. Maramura
Department of Public Administration & Management
University of the Free State
Bloemfontein, South Africa

Oratilwe Annastacia Khoza
Department of Public Administration & Management
University of the Free State
Qwaqwa, South Africa

ISSN 2524-5449 ISSN 2524-5457 (electronic)
Local and Urban Governance
ISBN 978-3-031-96972-0 ISBN 978-3-031-96973-7 (eBook)
https://doi.org/10.1007/978-3-031-96973-7

This Springer imprint is published by the registered company Springer Nature Switzerland AG
The registered company address is: Gewerbestrasse 11, 6330 Cham, Switzerland

Preface

This book was conceived from the critical concern of the complex challenges and evolving dynamics in urban water governance in postcolonial Zimbabwe. It discusses and offers comprehensive insights into the diverse issues surrounding water management, exploring the intersection of policy, technology, society, economy, institutions, resilience, and sustainability. The impetus for this book comes from witnessing the significant impact of inadequate water governance on communities, economies, and ecosystems. With Zimbabwe facing recurrent water crises characterized by shortages, pollution, and inefficient water management systems, the urgency for a systematic review and proactive strategies has never been greater. These challenges are exacerbated by political instability, economic hardships, and the adverse effects of climate change, making robust water governance a pivotal element of national security and public health.

The book is driven by the vision of achieving resilient and sustainable water governance in Zimbabwe. It aims to foster a deep understanding of the existing challenges and to catalyze action among policymakers, community leaders, and stakeholders to initiate meaningful reforms. The primary objective of this book is to provide a detailed analysis of the postcolonial state of water governance in Zimbabwe, drawing on empirical data, case studies, and expert opinions to map out the complexities involved. It aims to educate and inform a broad audience that includes students, academics, policymakers, and practitioners in the fields of environmental science, urban planning, and public policy.

Presenting a holistic view of the challenges and opportunities within Zimbabwe's urban water management systems, the book aims to:

- Elevate the discourse around urban water governance and its critical role in sustainable development.
- Propose strategic recommendations that are both practical and innovative, promoting long-term resilience and sustainability in urban water governance.
- Encourage active participation and collaboration among various sectors and communities to improve urban water governance.

- Inspire further research and dialogue on improving water management practices in Zimbabwe and similar contexts globally.

In pursuing these objectives, we were motivated by the urgent need to address the recurrent water crises in Zimbabwe's urban areas, crises that are emblematic of broader governance challenges within the region and similar settings globally. Throughout the research, we engaged with a range of stakeholders—from local community leaders and government officials to international experts in water governance. These interactions enriched our perspectives and deepened our understanding of the multifaceted nature of water issues in postcolonial Zimbabwe.

Our analysis draws on a resilience and sustainability framework to offer insights into how urban water systems can be managed to meet the needs of diverse populations while anticipating and adapting to future challenges. This approach recognizes the critical need for governance structures that are not only robust in the face of socioeconomic and environmental shifts but are also equitable and inclusive.

The book is structured to guide the reader through a logical progression of topics, starting with the foundational concepts of urban water governance (Chap. 2) and moving toward more specific discussions on historical legacies (Chaps. 1 and 2), policies (Chap. 3), institutions (Chap. 4), political economy and ecology (Chap. 5), climate change (Chap. 6), gender and water (Chap. 7), infrastructure (Chap. 8), gender and health (Chap. 9). The book concludes with a forward-looking perspective (Chap. 10), offering strategic recommendations and calling for ongoing research and policy refinement to adapt to changing circumstances and new challenges.

We hope that this book will serve as a valuable resource for those engaged in the vital task of managing Zimbabwe's water resources. May it inspire informed debate, catalyze meaningful action, and contribute to the development of effective, equitable, and sustainable water governance practices that will benefit all Zimbabweans. Welcome to a journey of understanding, challenge, and change in the realm of water governance!

Bloemfontein, South Africa — Abraham R. Matamanda
Bloemfontein, South Africa — Tazviona R. Gambe
Cape Town, South Africa — Johannes I. Bhanye
Bloemfontein, South Africa — Tafadzwa C. Maramura
Qwaqwa, South Africa — Oratilwe Anastacia Khoza

Acknowledgments

Writing this book has been a discovery, a challenge, and a profound learning experience. With deep gratitude, we acknowledge the contributions of many individuals and institutions whose insights, support, and dedication have made this work possible. First, we extend our heartfelt thanks to our academic colleagues at the University of the Free State, the University of Zimbabwe, and the University of Cape Town, whose rigorous debates and invaluable feedback sharpened the arguments presented in this book. Their expertise and willingness to share their knowledge have been indispensable.

We are particularly grateful to our research team, whose diligence and commitment to field research have provided the foundational data crucial for this book. Their tireless efforts in gathering information, conducting interviews, and navigating the complex terrains of urban Zimbabwe are deeply appreciated.

Special thanks go to the various nongovernmental organizations and community leaders in urban areas across Zimbabwe. Their on-the-ground insights and experiences have been instrumental in shaping the practical recommendations offered in this book. Their dedication to improving the water situation in their communities is not only inspiring but has also been a guiding light for this work.

We also acknowledge the financial support received from the Faculty of Economic and Management Sciences at the University of the Free State, without which the extensive research necessary for this book could not have been undertaken. Their funding and support of research initiatives that aim to solve real-world problems have been crucial to the success of this project. We appreciate the assistance with language editing and formatting of the book by Dorothea du Plessis.

We would also like to thank our families, who have always been there to support and encourage us. Their patience during the many hours we spent on this project have been pillars of strength for us.

Finally, we thank our editor and the publishing team at Springer. Their professionalism and attention to detail have greatly enhanced the quality of this publication, making the complex topic of water governance accessible and engaging for a wide audience.

To all who have been part of this journey—directly or indirectly—we are immensely thankful. Your contributions have not only enriched this book but have also deepened our commitment to advocating for sustainable and equitable water governance in Zimbabwe and beyond.

Competing Interests The author has no competing interests to declare that are relevant to the content of this manuscript.

Blurb About the Book

This book explores the critical and complex dynamics of urban water governance in Zimbabwe, a nation facing persistent challenges in this sector in the postcolonial era. Authored by renowned experts in urban planning, social ecology, public administration, and environmental policy—Abraham R Matamanda, Tazviona R Gambe, Johannes I Bhanye, Tafadzwa C Maramura, and Oratilwe Anastacia Khoza—the book presents the multifaceted aspects of urban water governance through a detailed analysis of the historical, political, and socioeconomic factors influencing urban water systems.

In Zimbabwe, urban water governance has reached a critical juncture. Frequent water shortages, deteriorating infrastructure, and inequitable water distribution are posing serious struggles for urban residents. This book explores these challenges through an interdisciplinary lens, blending insights from political science, public health, environmental studies, and urban planning to offer a comprehensive analysis of the current state and future prospects of water governance in Zimbabwe.

Each chapter of the book addresses different dimensions of urban water governance, from the theoretical underpinnings, colonial legacies, political instability, economic hardships, climate impacts, and legislative frameworks to the practical challenges of infrastructure financing and technological advancements. The authors critically examine how institutional fragmentation, the politicization of water resources, and inadequate infrastructure have led to significant urban water crises, which impact public health and exacerbate socioeconomic disparities. The book also scrutinizes the effectiveness of current policies and suggests comprehensive reforms aimed at enhancing the efficiency, equity, and sustainability of water governance systems.

The book sheds light on the complexities of water governance in a postcolonial context and also provides a pathway toward resilience and sustainability in urban water governance. It emphasizes the importance of integrating gender considerations, community engagement, and innovative financing mechanisms to enhance water security and governance. The comprehensive analysis presented in this book makes it an essential reading for policymakers, academics, and professionals in the fields of urban planning, geography, environmental science, and public policy. It also serves as a crucial resource for nongovernmental organizations, international aid organizations, and stakeholders involved in water governance and sustainable development initiatives, and anyone interested in understanding and solving the urban water challenges facing not only Zimbabwe but similar contexts globally.

Contents

About the Authors

Abraham R. Matamanda with a PhD in Urban and Regional Planning is also an NRF Y2-rated urban and regional planner. Abraham lectures at the University of the Free State (UFS) and is the Deputy Chair of the SARChI Chair on City–Region Economies in the faculty of Economic and Management Sciences at the UFS. Currently, he serves as the editor of the *Town Planning Journal* published by the UFS and also serves on the academic editorial board of the *PLOS Water* journal and sits on the Advisory Board of *Urban Research & Practice* Journal. He is a fellow of the third cohort of Phase I of the Future Professorate Programme at the Department of Higher Education and Training. His research focuses on urban governance and planning, climate change adaptation and resilience, informal Global South urbanism, urban land markets, and housing studies. Abraham has co-authored and co-edited several books published with Springer Nature, including:

- *Urban Geography in Postcolonial Zimbabwe: Paradigms and Perspectives for Sustainable Urban Planning and Governance*
- *Housing and Technology: Special Focus on Zimbabwe*
- *Urban Infrastructure in Zimbabwe: Departures, Divergences and Convergences*
- *Secondary Cities and Local Governance in Southern Africa*
- *Sustainable Development Goals and Urban Health: Strides, Challenges and Ways Forward for Poor Neighbourhoods*

Tazviona R. Gambe is a seasoned urban and regional planning specialist and researcher with over 17 years of experience. He holds formal training in rural and urban planning, real estate management, and monitoring and evaluation, which has equipped him to excel across industry, academia, and consultancy. In the real estate sector, he managed residential and commercial properties and conducted valuations for various assets, including residential, commercial, and industrial properties. Transitioning to academia, he has spent over a decade as a lecturer and researcher, focusing on African urbanization dynamics, regional economic resilience, urban water supply, and urban and regional development. He is currently a postdoctoral researcher at the University of the Free State, where he explores urbanization trajectories in African countries and the complex interplay between mining activities and urbanization on the continent. As a consultant, Tazviona has significantly contributed to spatial planning and environmental management. His notable projects include preparing master plans for the Masvingo rural district, Masvingo City, the Chiredzi rural district, and the Chitungwiza Municipality in Zimbabwe. He also played a pivotal role in the non-title survey of 126 low- and medium-density stands in the Gutu Rural District in Zimbabwe and conducted two environmental impact assessment studies. His career reflects a steadfast commitment to advancing sustainable urban and regional development in Africa through impactful research, teaching, and practical planning solutions.

Johannes I. Bhanye is an interdisciplinary researcher and urban planning scholar whose research revolves around four key themes: southern urbanism and urban informality; urban inclusion and exclusion; cities, social change, and transformation; and cities, space, and power. He is currently a postdoctoral research fellow at the African Climate and Development Initiative, University of Cape Town. His current research is on *Urban water adaptation: Exploring practices of adaptive governance for building urban water resilience in Cape Town's informal settlements*. Dr Bhanye holds a PhD in Social Sciences (Migration and Land Settlement), an MSc in Social Ecology, and a BSc in Urban and Regional Planning from the University of

Zimbabwe. He is also affiliated with the African Centre for Cities at the University of Cape Town, where he teaches the module *Climate Change and the City* for the master's program in Sustainable Urban Practice. Dr Bhanye is also a former research fellow at the University of the Free State, the Andrew W. Mellon Foundation, and the Centre for Applied Social Sciences at the University of Zimbabwe. He has published extensively in high-impact international journals such as *Urban Forum*, *Sustainable Development*, *Land Use Policy*, *Anthropology Southern Africa*, *PanAfrican Conversations*, *Journal of Asian and African Studies*, *Town and Regional Planning*, *African Identities*, and *International Journal of Sustainable Development & World Ecology*. His recent book *COVID-19 Lockdowns and the Urban Poor in Harare, Zimbabwe: Emerging Perspectives and the Morphing of a Sustainable Urban Future*, published by Springer. Outside academia, Dr Bhanye has more than seven years' experience in applied policy and development research, undertaking commissioned research and program evaluations and baseline studies on issues such as affordable housing, urban innovation, local economic development and inclusive urban economies, urban economic resilience, urban poverty and inequality (urban social geography), climate justice for marginalized populations, immigrants and racialized people's experiences for national and local government agencies, civil society, and international organizations such as USAID, Food and Agriculture Organization of the United Nations, CESVI Zimbabwe, Centre for Affordable Housing Finance Africa, and Homeless International. His research and professional experience spans across countries such as Zimbabwe, Senegal, Zambia, China, Germany, South Africa, Lesotho, Morocco, Belgium, Ghana, and Switzerland.

Tafadzwa C. Maramura graduated with her PhD in Public Management and Water Governance from the North-West University in South Africa and is currently a Senior Lecturer at the University of the Free State. As a Y-rated NRF researcher, her research interests span across sustainable service delivery, water governance, solid waste management, and green economies. She has published over 40 peer-reviewed research articles in both locally and internationally accredited journals, with findings recommended to countries such as South Africa, Zimbabwe, Botswana, Ghana, Kenya, and Rwanda. She is a Brightest Young Minds in Africa Alumni (BYM–2017) where she enjoys engaging with like-minded African think tanks.

Oratilwe Anastacia Khoza is currently a PhD candidate at the University of the Free State, specializing in the Water–Health nexus in informal settlements. She holds a master's degree in Administration, which she obtained *cum laude*. Her research focuses on critical areas such as water governance, sustainable service delivery, and the intersection of water and health challenges in informal settlements. With a strong academic background and passion for community-centered research, Ms Khoza has actively contributed to knowledge sharing and presented her findings at national and international conferences. Her work addresses societal issues by bridging the gap between water accessibility, health equity, and sustainable governance solutions.

Acronyms

Acronyms

AI	Artificial Intelligence
CCC	Citizen Coalition for Change
COZ	Constitution of Zimbabwe
CSOs	Civil Society Organizations
GIS	Geographic Information Systems
IoT	Internet of Things
IPCC	Intergovernmental Panel on Climate Change
IWRM	Integrated Water Resources Management
MDC	Movement for Democratic Change
NGO	Non-Governmental Organization
NUA	New Urban Agenda
NWP	National Water Policy
OECD	Organisation for Economic Co-operation and Development
SDG	Sustainable Development Goals
SIWI	Stockholm International Water Institute
SODIS	Solar Water Disinfection
UCA	Urban Council's Act
UNICEF	United Nations Children's Fund
US$	United States dollar
WHO	World Health Organization
WMO	World Meteorological Organization
ZANU–PF	Zimbabwe African National Union–Patriotic Front
ZINWA	Zimbabwe National Water Authority

Units of Measure

km	Kilometer
L	Liter
ML	Megaliter

Chapter 1
Introduction to Urban Water Governance in a Postcolonial Context

Abstract Water is recognized as a fundamental human right critical to sustaining life. In cities, the significance of water cannot be underestimated due to the increasing demand for water among different stakeholders. These stakeholders call for sound urban water governance, which enables sustainable allocation, access, and utilization of water among urbanites. However, the postcolonial urban context is marred by multiple issues compromising effective urban water governance. This introductory chapter thus sets the tone for the book and provides the background and context of urban water governance in postcolonial Zimbabwe. In the chapter, the introduction is presented, followed by the key arguments guiding the book, the aim and objectives and, lastly, the structure of the book.

Keywords Hydropolitics · Water governance · Urban governance · Sustainable development

1.1 Introduction and Background

Water is life. Since time immemorial, this precious liquid has been a critical resource in settlement planning, development and management. Access to adequate water is critical in sustaining the sustainability of settlements and cities as *"sufficient and reliable freshwater supply is essential for protecting human health"* (El-Fadel et al., 2003:107). This significance of reliable freshwater is evident from the case of London in the early 1800s, where the absence of freshwater in the city triggered a cholera epidemic. Between 1832 and 1866, at least 40,000 people succumbed to the cholera epidemic, which has been attributed largely to residents consuming contaminated water (Hinds, 2023:10). The need for adequate water resources to sustain humans has often shaped the form of human settlements. This has been identified since prehistoric times when settlements formed along river basins, for example, the settlements along rivers such as the Nile, Indus, Tigris, Euphrates, and Huang He (Macklin & Lewin, 2015; Phillipps et al., 2016; Wilson, 2012), or clustered along springs in deserts. In many cases, wars and conflicts have erupted as tribes,

A. R. Matamanda et al., *Urban Water Governance in Postcolonial Zimbabwe*, Local and Urban Governance, https://doi.org/10.1007/978-3-031-96973-7_1

communities, and nations sought to access and control water resources (El-Fadel et al., 2003; Petersen-Perlman et al., 2017).

Water is fundamentally recognized as a critical resource in cities as shown by the commitment to the Sustainable Development Goals (SDGs), specifically SDG 6, to ensure global citizens have access to water by 2030 (Mgwele et al., 2024). The same aspirations are shared by the African Agenda 2063, which envisages African cities that are modern, affordable, and livable by 2063. Specifically, the aspiration is that by 2063, African cities will be characterized by well-connected infrastructure that includes water and sanitation. Accordingly, the Africa Agenda of 2063 *"will be one where every citizen has affordable and sustainable access to quality basic services such as decent affordable housing, access to adequate and clean water and sanitation, transport and other services"* (Africa Union Commission, 2015:32). However, achieving this aspiration seems to be a top priority, considering that many African cities are currently grappling with water quantity and quality challenges (Balogun et al., 2017; Chitonge, 2014; Enqvist & Ziervogel, 2019). The situation is more critical in sub-Saharan Africa, where some communities, especially the poor neighborhoods, have had to endure weeks or even months without water running in the taps (Almer et al., 2017; Yates et al., 2022). Urban water supply has been one significant problem associated with rapid urbanization in many sub-Saharan African cities. Major cities such as Dar es Salam in Tanzania, Cape Town in South Africa, Lagos in Nigeria, and Kinshasa in the Democratic Republic of the Congo have not been spared. For example, the day zero experienced in Cape Town in 2018 reflected the severity of the water woes in southern Africa (Edmond, 2019; Joubert & Ziervogel, 2019). Some residents from Grasland in Bloemfontein, South Africa, are not connected to the reticulated water system, leaving the community to resort to unsafe water resources (Mphambukeli, 2019).

Failure in urban water governance has been attributed to water woes, especially considering that coastal cities have abundant water resources (Enqvist & Ziervogel, 2019; Okpanachi, 2008; Rugemalila & Gibbs, 2015). Scheba (2022) pointed to the strides made by the South African postapartheid government in addressing the urban water problems inherited from the apartheid government but points to several fissures and cracks relating to poor local governance. It has been noted that understanding urban water governance issues in the region (Southern Africa) thus requires a political ecology lens that facilitates understanding the deeper issues beyond the water crisis in cities (Millington & Scheba, 2021; Yates et al., 2022). For example, the Kingdom of Lesotho, which sits on massive water resource reserves, also experiences water insecurity problems, bringing to attention the politicization of urban water governance and State failure in managing urban water resources (Schwikoski, 2024). Therefore, studies are beginning to show that recurrent water crises in urban areas can be better understood within the context of the contentious and complex formal and informal institutions in water governance (Millington & Scheba, 2021; Scheba, 2022).

Rogers and Hall (2003:7) defined water governance as "the range of political, social, economic and administrative systems in place to develop and manage water resources, and deliver water services, at different societal levels". This definition brings out the complexity of water and reveals its interdisciplinary nature. Zimbabwe is no exception, as the country has been urbanizing since its independence in 1980. Although the larger share of the country's population still resides in rural areas (61.4% as indicated in the 2022 national census), it is projected that by 2040, half of the country's population will be residing in urban areas (Zimbabwe National Statistics Agency, 2022:11). At the same time, urban water insecurity has escalated, showing the mismatch between urbanization and urban infrastructure development (Bhanye et al., 2023). In Harare, many poor suburbs have been experiencing imminent water rationing since the early 2000s (Matamanda et al., 2024). Some emerging settlements, such as Hopley Farm Settlement, accommodating at least 7000 households, are not connected to the water supply grid (Ndoziya et al., 2019). While the problem may be attributed to climate change that has affected the hydrological cycle, it is also critical to note the influence of political and administrative issues on urban water challenges. Understanding the basis of these water insecurities is best done by exploring urban water governance in the postcolonial context of Zimbabwe. This exploration probes issues relating to hydropolitics, especially considering the regional disparities.

This book focuses on urban water governance in Zimbabwe and argues that the fragmentation of responsibilities between various institutions, the difficulty of coordination at various levels, and the politics centered on water constrain effective water governance in Zimbabwe. This has resulted in a downward spiral in urban service delivery in Zimbabwe's cities, towns, and smaller urban centers. Additionally, the innovative nature of this study is to draw from an often-neglected dimension in urban water governance, which is the political ecology which brings into perspective the varying interests associated with environmental contests and conflicts that influence how water resources are allocated and shared among urban areas by the public sector. While some environmentalists argue that climate change has a severe impact on urban water resources (Rugemalila & Gibbs, 2015), they would be oblivious to ignore the public management and governance of the urban water systems, its lack thereof and how it has exacerbated the water woes in Zimbabwe. Thus, understanding the politics, economics, and ecology of water governance through an interdisciplinary lens is crucial for this book because Zimbabwe's current state of obsolete water infrastructure has also had a negative impact on the supply of high-quality potable water services to the public in urban areas. This exploration probes issues relating to hydropolitics, geopolitics, and economic and environmental disparities, especially the motives and interests of the governance decisions made in managing urban water resources through the lens of political economy and ecology. Given this background, the book aims to explore urban water governance in postcolonial Zimbabwe from a resilience and sustainability perspective. The specific objectives of the book are the following:

- To dissect the extremities of institutionalism and urban water governance in Zimbabwe to gain co-creative insights that enable social change and resilience.
- To investigate the gender and urban water nexus challenges to provide a resilient and sustainable perspective for societal change.

- To identify the nature and extent of the political economy and ecology of urban water governance in a failed state to understand the impact it has on communities.
- To draw lessons for effective policies aimed at addressing urban water governance for sustainable development in urban areas.

1.2 Urbanization and Water Crisis in Zimbabwe

Before 1980, urban settlements in Zimbabwe emerged and existed as political and administrative structures that served the colonial government. These urban settlements were largely the preserve of the white population (Mpofu, 2011). The British colonial government enforced restrictive laws and policies that helped to keep most native Africans in the rural areas, thus ensuring that the urban population remained low. Legislation such as the Land Apportionment Act of 1930 and the Land Tenure Act of 1931 were instrumental in partitioning the country into European and African territories, where the latter occupied rural land and the former dominated the urban areas (Mseba, 2016). Africans could not own property in urban areas, and their presence in cities was mainly to provide cheap labor in the factories in major cities, including Harare, Bulawayo and Gweru (Bond, 1993). Munzwa and Wellington (2010) and Bond (1993) have reasoned that such restrictive laws imposed by the colonial government were meant to limit the urban population to maintain a balance between urban basic service demand and supply. Besides the segregatory nature of the urbanization that prevailed during the pre-independence era, the colonial government ensured that all urban residents had access to basic services, including water.

However, the situation changed in 1980 when Zimbabwe gained its independence from British colonial rule. The restrictive laws and policies were removed, and everyone had the right to the city. The result was an exodus of rural–urban migration as many Africans migrated to the urban areas in search of economic opportunities and to escape the poverty in the rural areas (Muchadenyika et al., 2018; Potts, 2006). Big cities such as Harare and Bulawayo attracted more people because of the presumed economic opportunities present in these larger cities. Subsequently, the urban population of the country rose from 23% in 1982 to 30% by the early 1990s, confirming the increasing population in the country's urban centers (Zimbabwe National Statistics Agency, 2012, 2016, 2022). Table 1.1 demonstrates the population increases experienced in the major towns and cities across Zimbabwe from 1980 to 2022.

Since the late 1990s, there has been an escalation of the water crisis in Zimbabwe's urban centers, including Harare, Bulawayo, Masvingo, Gweru, Murehwa, Chinhoyi and Chitungwiza (Davis & Hirji, 2014; Matsa et al., 2021; Musemwa, 2006; Zvobgo & Do, 2020). For Harare, the daily water demand for the city is estimated at 1,200,000 m^3. In contrast, the city produces approximately 6,000,000 m^3 per day,

Table 1.1 Population figures for selected towns and cities in Zimbabwe

Town/City	1982 Census	1992 Census	2002 Census	2012 Census	2022 Census
Harare	656,011	1,189,103	1,444,534	1,485,231	1,491,740
Bulawayo	413,814	621,742	676,787	653,337[a]	665,940
Chitungwiza	172,556	274,912	321,782	356,840	371,244
Mutare	69,621	131,367	170,106	187,621	224,802
Gweru	78,918	128,037	141,260	157,865	161,292
Kwekwe	47,607	75,425	93,072	100,900	119,863
Kadoma	44,613	67,750	76,173	92,469	117,380
Masvingo	30,523	51,743	69,993	87,886	90,286
Chinhoyi	24,322	43,054	56,794	77,929	103,671
Marondera	19,971	39,384	52,283	61,998	66,203

[a]Much debate has been raised with regard to this figure. Local authorities in Bulawayo argued that the population of the city was much more than the census figure, and they estimated the population of Bulawayo to be at least 1,1 million (Dube 2015; Mlotshwa 2012). This figure is based on the council's own study, which they termed the Consolidated Bulawayo Municipal Housing and Population Statistics. Source: Zimbabwe National Statistics Agency (2012, 2016, 2022).

almost half the demand (Davis & Hirji, 2014:22). The situation is equally dire for Bulawayo, which has been grappling with a water crisis since independence (Musemwa, 2006). Residents also extract water from informal sources, including rainwater harvesting, boreholes and unsafe surface water (Chirisa et al., 2014; Davis & Hirji, 2014; Gambe, 2019). However, these shallow wells are susceptible to sewage and other groundwater contaminations, eventually putting the residents at risk of waterborne diseases (see Fig. 1.1).

The water challenges being experienced in the towns and cities in Zimbabwe largely have to do with quantity and quality issues, as Makwara and Tavuyanago (2012,151) wrote that the quality and quantity of the water supply in most of these towns and cities have plummeted to crisis levels. This situation is clearly expressed in the following quote:

> *Water is a precious liquid and colourless. This definition always made sense for urban dwellers because they indeed always possessed a colourless precious liquid. Now they have a colourful but still very precious liquid* (Chirisa et al., 2014:72).

For example, many households in urban centers such as Harare, Bulawayo and Masvingo have gone for days, even months, without receiving water in their taps. In some instances where such situations have occurred, frequent pumping problems have been attributed to causing erratic water supplies. Examples are high-elevation areas such as Tafara and Mabvuku in Harare, where residents have gone for months with dry taps (Masiyiwa, 2020). Ironically, the water crisis is mainly experienced by the urban poor, who are excluded from the limited and reliable water infrastructure that was set up mainly for the rich during the colonial era.

Obsolete infrastructure, however, remains a key factor in the water crisis in most urban centers, considering that the current water infrastructure was installed in the 1930s by the colonial government. In addition to the ageing water infrastructure, it

Fig. 1.1 Shallow wells by Hopley residents; one is left open, and one has a lid to cover it; both are referred to as protected wells. (Source: Authors, 2018)

is also critical to note that the rapid urbanization experienced after independence has contributed to the massive increase in the urban population and the demand on water supply systems. As such, water shortages in Harare and other cities and towns are also caused by some informal and formal residential settlements being developed without connections to water mains. At times, these emerging human settlements are connected to the existing water infrastructure (Davis & Hirji, 2014). However, in most instances, the water system was designed for a small population; hence, it fails to sustain the increasing demand. Examples are drawn from Harare, where the current water infrastructure was meant to serve a population of 350,000, but at present, the same infrastructure now serves more than two million people. The same can be said for Bulawayo, where the water shortages are attributed to the water supply system that fails to match the current population (655,700 people) since it was meant to serve approximately 100,000 people (Moyo, 2013).

1.3 Urban Water Governance in Postcolonial Zimbabwe

While the foregoing provides a situational analysis of the water crisis in postcolonial Zimbabwe, we situate the debate and discussion in this book on urban water governance. This book focuses on urban water supply and governance in the Global South, using urban Zimbabwe as a case study to provide insights and perspectives into the realities of water governance. Applying a resilience and sustainability perspective, we argue that the fragmentation of responsibilities between various institutions, the difficulty of coordination at various levels, and the politics centered on water supply constrain effective water governance in Zimbabwe. This has resulted in a downward spiral in urban water services in Zimbabwe's urban centers.

Additionally, the innovative nature of this study is to draw from an often-neglected dimension in urban water governance, which is the political ecology that brings into perspective the varying interests associated with environmental contests and conflicts that influence water allocation and sharing among urban areas by the public sector. Therefore, this book dwells on understanding the politics, economics, and ecology of water governance through an interdisciplinary lens. This analysis is critical for this study because Zimbabwe's current state of obsolete water infrastructure has also had a negative impact on the supply of high-quality potable water services to the public in urban areas. This book will be critical for academics and professionals in the fields of urban geographers, planners, sociologists and water experts. It will also be useful for undergraduate and postgraduate students in geography, urban and regional planning, political science, development studies and economics.

1.4 Structure of the Book

This introductory chapter sets the tone for the book, focusing on the aim, objectives and argument by providing the background and context of urbanization and the water crisis in Zimbabwe, followed by the urban water governance in postcolonial Zimbabwe, and lastly, the structure of the book.

Chapter 2 presents the theoretical issues guiding urban water governance. The focus is specifically on the conceptual underpinnings of water governance, which helps to set the tone and foundation for subsequent chapters. The underlying issue and argument in the chapter is the recognition that water governance is complex and needs to be explored from a holistic perspective. The perspectives on water governance include definitions, typologies and principles of water governance. Water governance is linked to broader concepts of sustainability and resilience, which help to bring more insights into good water governance. Overall, the chapter argues that enhancing good urban water governance must be premised on the pillars of sustainability and resilience.

Chapter 3 examines the legislation and policy framework guiding urban water governance in postcolonial Zimbabwe. The analysis is enabled by a desktop review of key policies and acts of parliament that regulate and guide the governance of the urban water system in Zimbabwe. The ideologies informing the respective legislation and policies are examined to understand the motivation and expected outcomes, focusing on citizens' rights to adequate water and what it means. The chapter is framed in the context of distributive justice and articulates how policies and legislation are related to human needs and social justice. The findings from the analysis reveal that policies address multiple issues relating to water governance in urban areas.

Water as a resource has become a topical issue on the global agenda. Applying the new institutionalism theory, Chap. 4 reviews and critically assesses water institutions and their impact on water governance in urban Zimbabwe. The chapter reveals that institutions, ranging from government ministries and departments, local authorities, non-governmental organizations (NGOs), civil society organizations (CSOs), the private sector, and residents' associations, play a critical role in shaping water-related decision-making, water policies and driving behaviors related to water sharing and use. However, the fragmentation of responsibilities between various institutions, the difficulty of coordination at various levels, and the politics centered on water constrain effective water governance in Zimbabwe. This has resulted in a downward spiral in urban service delivery in the cities, towns and smaller urban centers in Zimbabwe. The chapter concludes that most urban water crises are institutional and governance-related, and efforts should be made to ensure capacity building, transparency, and collaboration of various institutions.

Chapter 5 explores the political economy and ecology of urban water governance in Zimbabwe. The chapter specifically questions the motives and interests of the governance decisions in managing urban water resources through the political economy and ecology lens. While acknowledging the governance initiatives adopted by the postcolonial government in water governance, for example, the Water Act and Zimbabwe National Water Authority (ZINWA) Act, the chapter highlights the vested political and economic interests of the decision-makers at the time. Understanding politics, economics, and ecology is crucial for this chapter because Zimbabwe's current state of obsolete water infrastructure has a negative impact on the supply of high-quality water services to the public in urban areas. The chapter suggests that to make sustainable political, economic, and ecological functionalities necessary, there is an urgent need to provide and suggest solutions for resolving these urban water governance difficulties.

Chapter 6 reveals that the impacts of climate change compound complex and multidimensional challenges for secure water service delivery in urban Zimbabwe, from supply shortages to falling groundwater levels. However, due to climate variability, urban residents apply various strategies in adapting to water shortages, ranging from reducing water use, using storage tanks, adopting climate-smart agriculture, water reuse and rainwater harvesting. While policy and institutional reform have advanced at the national level and on paper, nothing much has practically been done from the top to address the visible impacts of climate change on the water sector in urban Zimbabwe. The chapter concludes that adaptation to climate change should be at the center of urban water governance in Zimbabwe.

Chapter 7 argues that concerted efforts between men and women in addressing water woes in urban Zimbabwe are mutually beneficial. Most women and young girls are subjected to distressing experiences while gathering water for the household. Suburbs in various urban centers show some discrepancies as far as the burden of collecting water for the household is concerned. In low-income suburbs, women have a more significant role in managing household water issues. Yet, in the middle-income suburbs, women's involvement in gathering water for the household is less strenuous, especially in families with automobiles and supportive male counterparts. The chapter recommends a national quota system, stipulating that at least half of the members in any water committee or water department within municipalities should be females. This strategy can potentially increase women's participation in decisions that affect their daily lives, specifically water supply.

In Chap. 8, the authors focus on the financing of different types of water infrastructure and the implications for water governance. The chapter argues that an acute shortage of water infrastructure in Zimbabwe is one of the primary causes of water scarcity in most cities and towns. This chapter explores the innovative financing mechanisms and technological advancements that can help ease urban water shortages. The chapter found that the limited participation of the private sector in urban water supply inhibits private financing in water infrastructure. Although this has been an uphill task, central and local government institutions dominate the water supply sector and operate on a cost-recovery basis. Some urban centers have embarked on a drive to reduce the water purification cost to create extra water supply funds. The strategy involves synergies between cities and local companies working towards replacing imported with locally produced water purification chemicals. Similar synergies are recommended, especially in finding sustainable financing mechanisms that can improve investments in water infrastructure. Meanwhile, urban households should consider the uptake of low-cost household water filters that have become common in Africa's low-income communities. The filters can guarantee the consumption of clean and safe water.

Chapter 9 provides complex insights into how urban water governance challenges influence the water–health nexus because of the political instrumentalization of water as a human right. Concurrently, post-1980, the urban water infrastructural systems have increasingly depreciated, thereby leading to potable water challenges. In retrospect, this chapter highlights the water–health nexus because Zimbabwe has a history of frequent, severe cholera outbreaks that have resulted in at least 5000 deaths on average during every outbreak, the majority of whom fall into the category of people living in poverty. The cholera pandemic is a stark illustration of the unique problems underpinning urban water service delivery, which foster a favorable setting for a wide range of waterborne challenges. Furthermore, due to Zimbabwe's hyperinflationary economy, which obstructs low-income households from accessing alternative sources of potable water, people living in poverty are more prone to suffering from a wide range of water-related health diseases. As a result, the numerous health implications caused by fissures and cracks in the urban water governance structures must not be overlooked.

Lastly, Chap. 10 provides the lessons for urban water governance, sustainability and resilience in the postcolonial Zimbabwean context. It synthesizes the key themes discussed throughout the book, emphasizing the impacts of historical practices, various institutions, political dynamics, economic constraints, and technological advancements on water management. The chapter highlights the need for sustainable and resilient governance frameworks that are adaptable to both current challenges and future uncertainties, especially those posed by climate change and socioeconomic shifts. Reflecting on the complex impacts of inadequate water governance—from public health to environmental sustainability—the chapter proposes strategic recommendations aimed at enhancing the resilience, efficiency, and equity of water resource management in Zimbabwe. This synthesis not only addresses the immediate imperatives for improved water service delivery but also sets forth a vision for a water-secure future, advocating for cross-scalar governance and a unified approach that encompasses community involvement, technological innovation, and robust policy reforms.

References

African Union Commission. (2015). *Agenda 2063: The Africa we want*. African Union Commission.
Almer, C., Laurent-Lucchetti, J., & Oechslin, M. (2017). Water scarcity and rioting: Disaggregated evidence from Sub-Saharan Africa. *Journal of Environmental Economics and Management, 86*, 193–209. https://doi.org/10.1016/j.jeem.2017.06.002
Balogun, I. I., Sojobi, A. O., & Galkaye, E. (2017). Public water supply in Lagos State, Nigeria: Review of importance and challenges, status and concerns and pragmatic solutions. *Cogent Engineering, 4*(1), Article 1329776. https://doi.org/10.1080/23311916.2017.1329776
Bond, P. (1993). Economic origins of black townships in Zimbabwe: Contradictions of industrial and financial capital in the 1950s and 1960s. *Economic Geography, 69*(Issue 1: Theme Issue: African Development), 72–89. https://doi.org/10.2307/143890
Bhanye, J. I., Mangara, F., Matamanda, A. R., & Kachena, L. (2023). *COVID-19 lockdowns and the urban poor in Harare, Zimbabwe*. Springer. https://doi.org/10.1007/978-3-031-41669-9
Chirisa, I., Matamanda, A., & Bandauko, E. (2014). Ruralised urban areas vis-à-vis urbanised rural areas in Zimbabwe: Implications for spatial planning. *Regional Development Dialogue, 35*, 65–80.
Chitonge, H. (2014). Cities beyond networks: The status of water services for the urban poor in African cities. *African Studies, 73*(1), 58–83.
Davis, R., & Hirji, R. (2014). *Climate change and water resources planning, development and management in Zimbabwe: An issues paper*. World Bank. http://documents.worldbank.org/curated/en/254101468197068130
Edmond, C. (2019). Cape Town almost ran out of water. Here's how it averted the crisis. *World Economic Forum*. 23 August. https://www.weforum.org/agenda/2019/08/cape-town-was-90-days-away-from-running-out-of-water-heres-how-it-averted-the-crisis/
El-Fadel, M., El-Sayegh, Y., El-Fadl, K., & Khorbotly, D. (2003). The Nile River basin: A case study in surface water conflict resolution. *Journal of Natural Resources and Life Sciences Education, 32*, 107–117. https://www.agronomy.org/files/jnrlse/issues/2003/e02-15.pdf
Enqvist, J. P., & Ziervogel, G. (2019). Water governance and justice in Cape Town: An overview. *WIREs. Water, 6*(4), Article e1354. https://doi.org/10.1002/wat2.1354
Gambe, T. R. (2019). The gender dimensions of water poverty: Exploring water shortages in Chitungwiza. *Journal of Poverty, 23*(2), 105–122. https://doi.org/10.1080/10875549.2018.1517399
Hinds, N. (2023). *Circulations of waste in Victorian London*. (Doctoral thesis, University College London)
Joubert, L., & Ziervogel, G. (2019). *Day Zero: One city's response to a record-breaking drought*. Mapula Trust and African Climate and Development Initiative. http://dayzero.org.za/Day-Zero.pdf
Macklin, M. G., & Lewin, J. (2015). The rivers of civilization. *Quaternary Science Reviews, 114*, 228–244. https://doi.org/10.1016/j.quascirev.2015.02.004
Makwara, E. C., & Tavuyanago, B. (2012). Water woes in Zimbabwe's urban areas in the middist of plenty: 2000—present. *European Journal of Sustainable Development, 1*(2), 151–180. https://www.ecsdev.org/images/V1N2/makwara%20151-182.pdf
Masiyiwa, G. (2020). Residents of Zimbabwean suburbs scramble for water as city infrastructure falters. *Global Press Journal*. Online. https://globalpressjournal.com/africa/zimbabwe/residents-zimbabwe-suburbs-scramble-water-city-infrastructure-falters/. Accessed 14 October 2024

Matamanda, A. R., Mphambukeli, T. N., & Chirisa, I. (2024). Exploring water-gender-health nexus in human settlements in Hopley, Harare. *Cities & Health, 8*(1), 44–58. https://doi.org/10.1080/23748834.2022.2136557

Matsa, M., Mavugara, R., & Dzawanda, B. (2021). Urban domestic water supply crisis in the city of Gweru, Zimbabwe. *GeoJournal, 86*, 1173–1192. https://doi.org/10.1007/s10708-019-10118-x

Millington, N., & Scheba, S. (2021). Day Zero and the infrastructures of climate change: Water governance, inequality, and infrastructural politics in Cape Town's water crisis. *International Journal of Urban and Regional Research, 45*(1), 116–132. https://doi.org/10.1111/1468-2427.12899

Mgwele, A., Bhanye, J. I., & Mocwagae, K. S. (2024). Water and sanitation for urban health: A gender perspective on impacts and coping strategies in Mangaung Metropolitan Municipality. In A. R. Matamanda & V. Nel (Eds.), *Sustainable development goals and urban health: Strides, challenges and way forward for poor neighborhoods* (pp. 49–67). Springer. https://doi.org/10.1007/978-3-031-68734-1_4

Moyo, P. (2013). An assessment of private sector participation as a viable alternative for improved urban water provision in Zimbabwe: The Case of Harare Municipality (Master's thesis, Stellenbosch University, Stellenbosch, South Africa)

Mphambukeli, T. N. (2019). Social justice in planning: Access to adequate drinking water and contested belonging in emerging communities of Mangaung in post-apartheid South Africa. *African Renaissance, 16*(Special Issue 2) https://hdl.handle.net/10520/EJC-1a0a6686b0

Mpofu, B. (2011). 'Undesirable' Indians, Residential segregation and the ill-fated rise of the white 'housing covenanters' in Bulawayo, Colonial Zimbabwe, 1930–1973. *South African Historical Journal, 63*(4), 553–580. https://doi.org/10.1080/02582473.2011.626437

Mseba, A. (2016). Law, expertise, and settler conflicts over land in early colonial Zimbabwe, 1890–1923. *Environment and Planning A: Economy and Space, 48*(4), 665–680. https://doi.org/10.1177/0308518X15594807

Muchadenyika, D., Chakamba, M. K., & Mguni, P. (2018). Harare: Informality and urban citizenship –Housing struggles in Harare, Zimbabwe. In R. Rocco & J. van Ballegooijen (Eds.), *The Routledge handbook on informal urbanization* (pp. 124–134). Routledge.

Munzwa, K., & Wellington. (2010). Urban development in Zimbabwe: A human settlement perspective. *Theoretical and Empirical Researches in Urban Management, 5*(14), 120–146. http://www.jstor.org/stable/24861509

Musemwa, M. (2006). Disciplining a 'dissident' city: Hydropolitics in the city of Bulawayo, Matabeleland, Zimbabwe, 1980–1994. *Journal of Southern African Studies, 32*(2), 239–254. https://doi.org/10.1080/03057070600656119

Ndoziya, A. T., Hoko, Z., & Gumindoga, W. (2019). Assessment of the impact of pit latrines on groundwater contamination in Hopley Settlement, Harare, Zimbabwe. *Journal of Water Sanitation and Hygiene for Development, 9*(3), 464–476. https://doi.org/10.2166/washdev.2019.179

Okpanachi, E. (2008). Privatisation and universal access to water: Examining the recent phase of water governance in Nigeria. In J. In Feyen, K. Shannon, & M. Neville (Eds.), *Water and urban development paradigms: Towards an integration of engineering, design and management approaches* (pp. 637–644). CRC Press. https://doi.org/10.1201/9780203884102

Petersen-Perlman, J. D., Veilleux, J. C., & Wolf, A. T. (2017). International water conflict and cooperation: Challenges and opportunities. *Water International, 42*(2), 105–120. https://doi.org/10.1080/02508060.2017.1276041

Phillipps, R., Holdaway, S., Emmitt, J., & Wendrich, W. (2016). Variability in the Neolithic settlement patterns of the Egyptian Nile Valley. *African Archaeological Review, 33*, 277–295. https://doi.org/10.1007/s10437-016-9224-0

Potts, D. (2006). 'All my hopes and dreams are shattered': Urbanization and migrancy in an imploding African economy—The case of Zimbabwe. *Geoforum, 37*(4), 536–551. https://doi.org/10.1016/j.geoforum.2005.11.003

Rogers, P., & Hall, A. W. (2003). Effective water governance. *Global Water Partnership Technical Committee (TEC)*. https://www.gwp.org/globalassets/global/toolbox/publications/background-papers/07-effective-water-governance-2003-english.pdf

Rugemalila, R., & Gibbs, L. (2015). Urban water governance failure and local strategies for overcoming water shortages in Dar es Salaam, Tanzania. *Environment and Planning C: Politics and Spaces, 33*(2), 412–427. https://doi.org/10.1068/c1324

Scheba, S. (2022). Viewpoint—The South African water sector: Municipal dysfunction, resistance and future pathways. *Water Alternatives, 15*(3), 632–649. https://www.water-alternatives.org/index.php/alldoc/articles/vol15/v15issue3/676-a15-3-5/file

Schwikowski, M. (2024). Lesotho's water: will villagers be compensated?. https://www.dw.com/en/lesothos-water-will-villagers-be-compensated/a-69352942

Wilson, P. (2012). Waterways, settlements and shifting power in the northern-western Nile Delta. *Water History, 4*(1), 95–117. https://doi.org/10.1007/s12685-012-0053-z

Yates, J. S., Tadaki, M., & Clark, C. (2022). Political ecologies of urban water governance. In T. Bolognesi, F. S. Pinto, & M. Farrelly (Eds.), *Routledge handbook of urban water governance* (pp. 331–344).

Zimbabwe National Statistics Agency. (2012). *Zimbabwe population census 2012*. Zimbabwe National Statistics Agency.

Zimbabwe National Statistics Agency. (2016). *Facts and Figures 2016*. Zimbabwe National Statistics Agency.

Zimbabwe National Statistics Agency. (2022). *2022 Population and housing census: Preliminary report on population figures*. Zimbabwe National Statistics Agency.

Zvobgo, L., & Do, P. (2020). COVID-19 and the call for 'Safe Hands': Challenges facing the under-resourced municipalities that lack potable water access—A case study of Chitungwiza municipality, Zimbabwe. *Water Research X, 9*, Article 100074. https://doi.org/10.1016/j.wroa.2020.100074

Chapter 2
Conceptual Underpinnings on Urban Water Governance

Abstract This chapter provides the conceptual and theoretical arguments for urban water governance. It recognizes that water governance is a complex process that requires governments to uphold the social, economic, environmental and political issues inherent in water governance processes. The dimensions of water governance illustrate the need to adopt a holistic perspective through which water governance is conceptualized as a continuously evolving process that ought to satisfy the fundamental water rights of all without marginalizing them. The types of water governance are discussed with a focus on authoritarian and participatory water governance processes. The chapter further explains the principles of good governance, which help to determine the success or failure of urban water governance. The chapter ends with theoretical discussions on resilience and sustainability, which are critical in spurring the effectiveness and robustness of urban water governance processes.

Keywords Resilience · Sustainability · Stakeholder engagement · Complexity

2.1 Introduction

This theoretical chapter considers the nuances and intricacies of urban water governance. The focus is on the conceptual underpinnings of water governance, which is the foundation of the consecutive chapters presented in this book. The chapter argues that urban water governance is a complex global issue that governments grapple with. The chapter critically asks the pertinent question: What is urban water governance? While attempting to demystify the concept *urban water governance*, the chapter recognizes critical issues relating to urban water governance. These include articulating and discussing the dimensions, typologies, and principles of water governance. In articulating these issues, the chapter considers the notion of good water governance, which is encouraged as it helps alleviate the urban water challenges, such as overcoming quality and quantity deficiencies. Stakeholder engagement is also vital, including considering the concerns of disadvantaged individuals and groups. The chapter then discusses the theoretical underpinnings and

A. R. Matamanda et al., *Urban Water Governance in Postcolonial Zimbabwe*, Local and Urban Governance, https://doi.org/10.1007/978-3-031-96973-7_2

principles related to water governance. First, the notion of resilience is critical in understanding water governance as this brings to attention the ability of the water system to withstand several shocks inherent in the urban water system. Additionally, resilience is about the process of governance tailor-made to local conditions, thus fused into the governance process and system. Second, sustainability is discussed concerning water governance. Sustainability embodies bringing in socioeconomic and environmental concerns in water governance to ensure equity in how urban water is governed. The chapter argues that enhancing good urban water governance must be premised on the pillars of resilience and sustainability.

2.2 Defining Water Governance

Water governance is a concept that gained popularity as a policy concern in recent studies in response to increasing perceptions of water crisis, including climate change and other competing uses (Akamani, 2016; Castro, 2007; Özerol et al., 2018; Woodhouse & Muller, 2017). Zwarteveen et al. (2017:3) defined water governance as "the practices of coordination and decision-making between different actors around contested water distributions." In this definition, water governance is considered as a practice of mediating contested uses and claims around water use and access, which thus calls for coordinated practices and decision-making to regularize uses and manage water distribution. The definition also somehow recognizes power dynamics because contestations are largely centered around conflicts that are associated with scarce resources; hence, those with power and authority influence the decision-making process and the allocation of the water resources. Schultz (2019:2) defined water governance as "a combination of (a) water policy (the institutional framework); (b) water politics (power relations between political actors); and (c) water policy (the mechanisms and instruments used to achieve certain outcomes)".

Similarly, Jiménez et al. (2020:3) defined water governance as "a combination of functions, performed with certain attributes, to achieve one or more desired outcomes, all shaped by the values and aspirations of individuals and organisations". This book will adopt the definition proffered by Schultz (2019:2) who has defined water governance as "a combination of (a) water policy (the institutional framework); (b) water politics (power relations between political actors); and (c) water policy (the mechanisms and instruments used to achieve certain outcomes)". Essentially, this definition covers all facets of the concept, including social, political, environmental and economic systems of water governance.

Various urban water governance regimes around the world have transformed to facilitate the development and management of water resources as well as water services supply (Özerol et al., 2018). First, the idea of water governance incorporates the capacity to develop institutional frameworks and public policies that are socially accepted and actuate resources that support them. Second, water governance highlights the administrative and political fundamentals of dealing with a problem or abusing an opportunity. Third, water governance is interested in decision-making processes that take place at institutions and involve various organizations, stakeholders or actors that are of the essence for water management and water development (Lautze et al. 2011; Rogers& Hall, 2003; Schulz, 2019). For good governance of water, a combination of committed governments from local to national, along with the private sector, NGOs and other groups in civil society is necessary to address governance-related values that include sustainability, economic efficiency, and social justice (Rogers & Hall, 2003; Schulz, 2019).

2.3 Water Governance Dimensions

Water governance is a complex issue. Four dimensions and perspectives are critical to consider in water governance, namely social, economic, political and environmental dimensions (Stockholm International Water Institute [SIWI], 2016, 2024). First, **society** is often at the center of water governance. This is largely due to the diversity of society, with individuals and groups coming from different backgrounds and with varying social norms, values and aspirations. For a long time, especially in colonial cities, resources have been allocated along racial lines where the white minority have tended to receive better quality services, while Africans received poor quality services, usually accessing water from communal taps. This issue highlights the perspectives on equitable access to water resources, which resonates with the fundamentals of social injustices (explained in Chap. 3).

2.3.1 Social Dimension

The **social dimension** of water governance advocates for the equitable distribution of water resources and services among different economic and social groups and its impact on society (SIWI 2016, 2024). Apart from the fact that water resources are unevenly distributed in space and time, they are also unevenly distributed between many socioeconomic groups in both urban and rural settlements. Mokoena (2022) showed how residents in Khayelitsha, an informal settlement in Cape Town, South Africa, are disadvantaged in accessing potable water. This marginalization of residents in informal settlements is a common occurrence evident in most (Southern) African cities where low-income households and those residing in informal settlements are least prioritized in water access (Mphambukeli 2019). The prevailing situation shows inequalities along income lines, whereas in the past, during the colonial era, it was mainly along racial lines.

2.3.2 Economic Dimension

Second, the **economic dimension** of water governance focuses on efficiency in water use and allocation and the role that water plays in overall economic growth (SIWI 2016, 2024). Freshwater resources are always in great demand and contested, especially in rapidly urbanizing regions where urbanites require potable water supply. For example, in Iran, the demand for the country's freshwater resources has been soaring and the multiple stakeholders contesting to access the water increases the value of the water resources, which for a long time has necessitated the need for the reformed governance of groundwater resources in the country (Sarami-Foroushani et al. 2023). In southern Africa, the government of Lesotho has also leveraged its economy on the abundant freshwater resources in the country, which is mostly sold to South Africa. In 2024, the government of Lesotho agreed with South Africa on an annual agreement of US$200 million for water (Lombard 2024). This amount shows the significant gains the government of Lesotho benefits from its water resources, duped *white gold*. While the government of Lesotho has managed to make some significant economic gains from exporting water to South Africa, some communities in Lesotho have been grappling with water access, demonstrating the government's failure to recognize the water needs of its local communities before exporting (Mokhethi and Kabi 2021; Molupe 2023). Therefore, water can be considered as an economic resource capable of supporting local economies with a high dependency on water and other natural resources. In this regard, water resources, especially groundwater, can be used for effective economic growth and poverty alleviation. Governance structures exert a powerful effect on per capita income in numerous countries.

The economic dimension of water is also highlighted in how water permits can be provided for water access. This is a common practice in most African countries where water permits are provided to users to access water services. For example, in South Africa, the National Water Act 36 of 1998 stipulates that water use licenses should be acquired for specific water uses, including water storage, water abstraction and using water for recreational purposes. Likewise, the government of Zimbabwe also mandates individuals and companies to apply for water permits when drilling boreholes and accessing water from natural reservoirs for personal gain. Water supplied to households and other stakeholders in cities should be paid for to enable the local authorities to sustain the water system. However, water pricing at times is not justified as households are charged more, especially when the water sector is privatized, and eventually, private companies are profit-oriented and seek to maximize profits from the water supply services they provide.

2.3.3 *Political Dimension*

Third, the **political dimension** of water governance represents equal rights and opportunities for stakeholders in water governance to contribute to decision-making processes. Participation regulates and facilitates more informed decision-making, improves conflict resolutions and encourages more effective implementation (Fainstein and Lubinsky 2020). Tyhotyholo and Ncube (2023) highlighted that a lack of community participation is one of the major challenges facing effective water governance in South Africa. Therefore, more effective involvement of marginalised residents, including slum dwellers, Indigenous people and the acknowledgement of those marginalised residents as legitimate stakeholders in water-related decision-making, stands to significantly enhance outcomes.

2.3.4 *Environmental Dimension*

Fourth, the **environmental dimension** of water governance is about the sustainable use of water and other related ecosystem services (SIWI 2024). The adequate flow of water at suitable quality is crucial to sustaining ecosystem functions and services that build upon them. Sadly, water quality is facing a serious decline in numerous regions because of inadequate safeguards in the way it is used by intensive agriculture and in large industrial and urban areas. In Lesotho, the mining activities at the Letšeng Diamond Mine have been attributed to the pollution of the Khubelu River, which is a major source of water supply in the country, thereby jeopardizing the health of the communities depending on this water supply (Kabi 2022). Likewise, in Zimbabwe, the City of Harare has been accused of discharging raw sewage in Lake Chivero, one of the water sources for Harare, placing communities at risk of waterborne diseases.

2.4 Types of Urban Water Governance

There are different types of urban governance. First, participatory urban water governance is a common approach in many recent studies relating to water governance. Participation has formed part of the "standard repertoire" of the concept of urban water governance (Van Buuren et al. 2019:369). Participation includes the engagement or consultation of stakeholders in decision-making processes while empowering the local communities and stakeholders to involve themselves in solving social challenges (Van Buuren et al. 2019). Participatory governance was shaped by ineffective state performance and more effective creativities of civil society and private sectors to tackle sociopolitical difficulties (Rana and Piracha 2020). Successful participatory governance can lead to long-term involvement of stakeholders and communities in improving the provision and administration of public goods and services. It can also open a new door for all members of society to be included in social well-being discussions where all their needs are taken into account to enhance and sustain their quality of life and well-being (Morinville and Harris 2014; Rana and Piracha 2020).

2.4.1 Participatory Water Governance

Participatory water governance is about public participation and the involvement of stakeholders and private sectors in decision-making processes to deal with issues related to water management and the provision of water services and resources. Common features of good water management include participation and deconcentrated transparency, decision-making and accountability (Grassini 2019). There are different motives for making efforts towards participatory water governance, including "increasing support for policies, enhancing problem-solving capacity, breaking deadlocks, avoiding litigation costs or improving the quality of decisions" (Van Buuren et al. 2019:369). As a result, participatory water governance was a crucial feature in all significant policy documents produced since the late 1990s in both developed and developing countries (Grassini, 2019). It also promotes collaborative governance where actors outside public agencies, levels of government and private or public spheres carry out a public purpose that could not otherwise have been accomplished (Emerson and Nabatchi 2015; see also Bruun and Rubin, 2022:539; Harrington, 2017).

However, Harris and Morinville (2013) argued that even though participation has multiple benefits on various levels, cautiousness is equally necessary. For example, participation can serve as a continuous responsibility to communities or certain groups in communities, including people living in poverty or women when there are insufficient resources and capacity. Moreover, delegation of power from central to local government for governance responsibilities can result in resources captured by the local elites or participation that is not fruitful; thus, deep-rooted power dynamics, which continues the marginalization of poor populations (Harris and Morinville 2013). However, the consequences of decentralization without coordination undermine the predicted performances of water governance and create a set of divided power blocks.

In addition, stakeholder involvement dives deeper into numerous methods of stakeholder involvement, such as bottom-up and top-down approaches (Van Buuren et al. 2019). The approach that has been on the rise in water governance and management is a bottom-up participatory approach. When faced with predicaments that have an impact on their interests, local actors, including the general public, NGOs, stakeholder groups and private companies, arrange to develop a form of collaborative action. These initiatives can emerge from disappointments or dissatisfactions that local actors have from efforts made by government-led participation as their response. Some examples of initiatives created for community-based initiatives and collaboration in water governance are the teamwork energies of local stakeholders to preserve and protect nature while managing fishing and tourism (Eckerberg et al., 2015). The top-down approach is most commonly found when discussing authoritarian water governance.

2.4.2 Authoritarian Water Governance

The term 'authoritarian' is a concept that has different meanings; however, this paper used the definition of Hegazi (2019:7) who defined authoritarian systems as "those that lack the qualities of a democratic system, and early definitions of non-democracies focused on the extent of political diversity and mobilization, and the presence, or lack thereof, of ideology." This means that the governance under authoritarianism does not give the public a platform during decision-making processes to express their opinions on the societal challenges that they experience. Environmental governance is a broader perspective of water governance, and the definition of authoritarian environmental governance is defined by two aspects (Lo, 2015). First, the policy procedures are ruled by a self-governing state where the policy process does not leave room for public participation. Second, environmental outcomes are pursued by limiting an individual's quality of freedom. For water governance, authoritarian governance prevents communities and stakeholders from participating in the policymaking and decision-making processes of water management and governance. Participatory governance is characterized by the bottom-up participatory approach, whereas authoritarian governance is primarily guided by a top-down approach.

The top-down participatory approach, also known as government-led participation in water management and governance, causes a decline in public participation because it operates under the assumption that those in control know what is best for everyone else (Maher and Buhmann 2019). The prospects of influence are expressed, but they are usually not met in practice (Van Buuren et al. 2019). As a result, societal actors are dissatisfied and unhappy with the rules of engagement, especially when participation in decision-making is limited.

Bruun and Rubin (2023) contested that even though collaborative water governance has been receiving traction as a response to water crisis management, authoritarian political structures make collaborative governance bear trajectories of one-party rule. Instead of using collaborative governance as a means of pursuing trust-based knowledge sharing, agreements, and compromises, it mainly serves the purpose of building legitimacy and public support through enhanced administrative and technical competence in implementing centrally defined policies (Bruun and Rubin, 2023). Equally, instead of encouraging purposeful discussions of issues and replies in the quest for bottom-up accountability, collaboration is mostly directed towards problems, processes, and contexts that do not contest the allocation of resources and power in society. Therefore, Bruun and Rubin (2023) argued that authoritarianism is not a suitable way of governing water because water governance is reliant on horizontal systems of actors that participate in repeated interactions at several levels.

Yan et al. (2022) emphasized that collaborative water governance is more than addressing neutral managerial and technical issues. Instead, it is inherently a political process that shows a complex web of power, narratives and interests. In theory, the expected benefits of collaborative water governance include enhanced participation, better responsiveness and efficiency, good conflict management, and enhanced capacity to address complex issues of water management (Harrington 2017; Yan et al. 2022). However, in reality, the successful implementation processes of collaborative water governance depend on the presence of participation, transparency, trust, and accountability, which are mostly absent in other authoritarian systems. Differences in collaborative water governance in various political systems and their performance gap have been explored in developing countries as well. Work by Guidone (2019) on Sekhukhune mining practices and water governance in South Africa highlighted that corruption and collusion of government representatives, mining companies, and traditional leaders (poor water governance) negatively affect the quality and availability of clean water for local societies.

2.5 Principles of Good Urban Water Governance

In 2009, the Organisation for Economic Co-operation and Development (OECD) established a water governance program to identify and help governments at all levels, to link critical governance gaps in the design and implementation of their existing water policies, through policy conversations, economic analysis, standards and international best practices (Akhmouch and Correia 2016; Akhmouch et al. 2020). This program reached its fullest potential when the OECD principles on water government were developed and adopted in 2015 through a multi-stakeholder and bottom-up process under the OECD water governance initiative (Akhmouch et al. 2020). These principles were expressed through three harmonizing and mutually strengthening dimensions of water governance, namely effectiveness, efficiency, and trust and engagement.

First, effectiveness is concerned with the influence of governance to define clear sustainable water policy goals and directs all spheres of government, to implement those policy goals and to meetall expected targets (Akhmouch and Correia 2016; OECD, 2015). It calls for clearly distributing roles and responsibilities, managing water at suitable scales, inspiring policy coherence, and adapting the level of capacity to the intricacy of water challenges to be met. When enhancing the effectiveness of water governance, four principles are considered (Akhmouch et al. 2020; OECD, 2015). Principle 1 is about clearly allocating and distinguishing roles and responsibilities for water policymaking, policy implementation, operational management and regulation, and fostering coordination across these responsible authorities. Principle 2 looks at managing water at suitable scales within integrated basin governance systems to reflect local conditions and encourage the development of coordination among the various scales. Principle 3 encourages policy coherence, especially considering the complexity of urban water governance policies. Lastly, Principle 4 considers adapting the capacity and responsibilities of authorities to engage with the complex challenges posed by water issues.

Second, efficiency is about the contribution of governance to exploit the advantages of sustainable water management and welfare at the lowest rate to society (Akhmouch et al. 2020). The OECD (2015) stated that the efficiency of water governance depends on four principles: Principle 1 is related to manufacturing, updating, timely distributing of consistent, comparable data and information related to water policy and water, to utilize it to direct, assess and improve water policy. Principle 2 is around ensuring that the governance preparations assist to mobilize water finance and distribute financial resources in an efficient, transparent and timely way. Moreover, Principle 3 ensures that regulatory frameworks for sound water management are successfully implemented and enforced in the public interest. Lastly, Principle 4 focuses on promoting the adoption and implementation of innovative water governance practices across accountable authorities, levels of government and all relevant stakeholders.

Third, trust and engagement relate to the involvement of governance in building public confidence and guaranteeing the inclusiveness of stakeholders through democratic legitimacy and fairness for all societies across all stages. It is about mainstreaming integrity and transparency, managing trade-offs across water users, urban and rural areas, and generations, and promoting regular monitoring and evaluation to make adjustments when needed (OECD 2015). Trust and efficiency, like the other dimensions listed above, are based on four water governance principles. Principle 1 focuses on mainstreaming transparency and integrity applications across water policies, water institutions, and water governance frameworks for better responsibility and trust in decision-making. Principle 2 of trust and engagement looks at promoting stakeholder engagement for informed and outcome-oriented contributions to water policy design and implementation. Principle 3 inspires water governance frameworks that assist in managing trade-offs across water users, urban and rural areas and generations. Principle 4 promotes consistent monitoring and evaluation of water policy and governance, and where necessary, sharing the results with the public and making adjustments when needed.

2.5.1 Resilience Theory and Urban Water Governance

The resilience theory was developed in the 1970s by Crawford S. Holling, head of the Ecological Policy Group, along with a group of diverse scientists at the University of British Columbia, with the target of reconstructing the face of ecology and natural resource management (Curtin and Parker 2014; Walker and Cooper 2018). The theory was pioneered to work on breaking down intellectual silos, making it more available to practitioners, and developing a nexus of social and ecological perspectives in a holistic approach to multifaceted issues that used large-scale studies as theoretical experiments (Curtin and Parker 2014). In his classic work *Resilience and Stability of Ecological Systems*, Holling (1973) demonstrated the destabilization of the notion of *equilibrium* as the core of the ecosystem concept and regular boundaries of ecosystem trajectories (see also Burnside et al. 2022; Walker and Cooper 2018). It contested the idea that there is an existing balance of nature where ecosystems will return to their original self if left to self-repair. After the concept of resilience grew in the field of ecology, it further infiltrated areas of social sciences, including finance, psychology, urban planning, development, corporate strategy, education, central banking, engineering, national security, psychiatry and public health (Walker and Cooper 2018).

The concept of *resilience* possesses various definitions as it is adopted in different fields of study. The original definition of resilience can be attributed to Holling (1973), who defined it as a "measure of the persistence of systems and of their ability to absorb change and disturbances and still maintain the same relationships between populations and state variables". Fiksel (2006:16) defined resilience as "the capacity of a system to survive, adapt and grow in the face of change and uncertainty". In its Presidential Policy Directive-8, The White House (2011:6) defines resilience as referring "to the ability to adapt to changing conditions and withstand and rapidly recover from disruption due to emergencies". Brand and Jax (2007) also addressed various definitions of resilience in sustainability, social-ecological and sociological perspectives. Resilience of water systems is analyzed through social-ecological systems (Saikia and Jiménez 2023).

Recognizing that water resource management and water services are systems is an important aspect to consider in building resilience (Saikia and Jiménez 2023). The social-ecological resilience is defined as "the capacity … to absorb, cope with, and adjust to changing social or environmental conditions while retaining key functions, structures, and identity, including people's capacity to learn and engage in shaping transformative change" (Fallon et al. 2022:3). Simultaneously, it is important to comprehend social-ecological systems as they are closely connected to water resilience. *Water resilience* is the concept adapted to water systems and water governance, more specifically from drought and flood management to climate change adaptation in the water services sector, as well as watershed and catchment-scale water resource management (Baird et al. 2016; Rijke et al. 2013; Rodina 2019; Salinas Rodriguez et al. 2014). Social-ecological systems and water resilience are connected because water systems themselves are placed at the interface between social and ecological systems. Water systems are both natural and social, calling for interdisciplinary approaches to study interactions of social (humans) and ecological (biophysical systems, ecosystems and the environment) that are influenced by internal and external forces (Saikia and Jiménez 2023).

Therefore, the important step of constructing water system resilience is to first identify and comprehend a specific system of social-ecological procedures that impact water supply and water access before working on collectively building resilience. Rockström et al. (2014), in their book *Water Resilience for Human Prosperity*, were the first to centralize governance and water resource management with an effort to achieve social-ecological resilience and sustainability. Water resilience in their context was different because it is incorporated in water governance where the meaning and implication of resilience changed.

Another argument around water resilience was made by Falkenmark et al. (2019), who insisted that water resilience is not only a matter of water but focuses on the role that water plays in the resilience of societies and ecosystems. Moreover, the concept of water resilience is diverse and focuses on various water sectors and dimensions of risks, therefore creating challenges for conceptual and practical water governance. As a result, there is no single way of defining water-related resilience because it contributes to improved water security for climate change on the one hand and decreases vulnerability to water-related risks and hazards on the other (Rodina 2019). For specific water resilience dimensions, there are more arguments on increased flexibility in the water sector and for increased reliance on natural or green infrastructure as a flexible option with multiple benefits, from improving water quality and supply to flood risk mitigation (Rodina 2019).

The resilience theory further incorporates three core capacities that are beyond the typical understanding of resilience as a conservative concept, namely absorptive, adaptive, and transformative capacity (Fallon et al. 2022). However, for water resilience, the principles of adaptiveness and flexibility are necessary in the context of water governance (Rodina 2019). Adaptive governance is a key emerging concept in the literature concerning water resilience. Adaptive governance has been employed to define a governance process that replies to feedback received by a managing agency, for example, adaptive management, from monitoring the response of the ecosystem (Cosens and Williams 2012). Adaptative governance has also been developed as a key analytical approach to understanding natural resource governance (Saikia and Jiménez, 2023).

2.5.2 Sustainability and Urban Water Governance

Sustainability is a normative concept that refers to physical and institutional practices that meet the needs of the present without compromising the ability of future generations to meet their own needs. For a system to be perceived as *sustainable*, a subtle balance and interplay between the requirements of the social, economic and environmental must be achieved. In the same breath, sustainable development challenges in water services, specifically management, governance and technical solutions, have been receiving increased attention over the last four decades (Di Vaio et al. 2021). Internationally recognized organizations such as the OECD and the United Nations Environmental Programme have emphasized the important role that water governance plays in sustainability issues (Di Vaio et al. 2021). Water sustainability issues have also been incorporated in the United Nations 2030 Agenda, which provides a framework for measuring sustainable development from a social, environmental and economic perspective through its 17 SDGs (Woodhouse and Muller 2017). On that note, water governance has been recognized as an important aspect in sustained progress towards achieving SDG 6, which aims to ensure sustainable management and availability of water and sanitation for all.

Various sustainable urban water governance frameworks and models exist as a way of achieving SDG6, as well as evaluating the sustainability of urban water governance. However, Schneider et al (2015) and Wiek and Larson (2012) contested that even though many research studies are focused on problems of more sustainable water governance, there are limited holistic approaches that evaluate water governance sustainability based on comprehensive, interdisciplinary research. Most frameworks emphasise singular aspects of water governance. For instance, a study by Hill (2012) has developed a framework on water governance and adaptive capacity, while Pahl-Wostl (2006) and Pahl-Wostl et al. (2007) have developed frameworks from water governance and social learning.

Holistic framework approaches of water governance include one by Wiek and Larson (2012), where their framework links the systematic understanding of water governance systems and its evaluation through seven sustainability principles. These principles are as follows: social-ecological system integrity, resource efficiency and maintenance, livelihood sufficiency and economic opportunity, social-ecological civility and democratic governance, intergenerational and intragenerational equity, interconnectivity from local to global scales, and precaution (mitigation) and adaptivity (Wiek and Larson 2012). The principles also clearly detail the concept of sustainable water governance along with other general sets of sustainability criteria applied to water sustainability, water sustainability proposals, water governance and natural resource governance.

Another holistic framework was proposed by Schneider et al. (2015) through the development of a Sustainability Wheel. The generation of four essential principles of sustainable urban water governance was the first step in developing the Sustainability Wheel. The principles include the contribution to societal goals of regional development, maintenance of ecological and hydrological integrity, contribution to social justice, and adaptive capacity (Schneider et al. 2015). The principles further allowed the development of indicators that ultimately created a full 'Wheel of Sustainability'. Also, sustainability is now recognized as the widely used framework in natural resource management, and it is sometimes represented as a triangular model that balances three competing interests of the economics, society and the environment (equitable access and cost, resource conservation and efficient supply).

Furthermore, other studies identified key elements that water governance needs to be sustainable and resilient, namely coordination, cooperation and collaboration, stakeholder engagement, adaptiveness, and stakeholder participation (Di Vaio et al. 2021). These key elements refer to different ways that actors should interact to achieve the common goal of ensuring water accessibility and sanitation for all. Di Vaio et al. (2021) further emphasized that these key elements form part of good governance for water systems and management. Ultimately, the sustainability paradigm necessitates multidisciplinary action and the involvement of various stakeholders from all levels of the system.

References

Akamani, K. (2016). Adaptive water governance: integrating the human dimensions into water resource governance. *Journal of Contemporary Water Research & Education, 158*(1), 2–18. https://doi.org/10.1111/j.1936-704X.2016.03215.x

Akhmouch, A., Clavreul, D., Hendry, S., Megdal, S. B., Nickum, J. E., Nunes-Correia, F., & Ross, A. (2020). Introduction: introducing the OECD principles on water governance. In A. Akhmouch, D. Clavreul, S. Hendry, S. B. Megdal, J. E. Nickum, F. Nunes-Correia, & A. Ross (Eds.), *OECD principles on water governance. From policy standards to practice* (pp. 5–12). Routledge.

Akhmouch, A., & Correia, F. N. (2016). The 12 OECD principles on water governance–When science meets policy. *Utilities Policy, 43*(Part A), 14–20. https://doi.org/10.1016/j.jup.2016.06.004

Baird, J., Plummer, R., Moore, M.-L., & Brandes, O. (2016). Introducing resilience practice to watershed groups: What are the learning effects? *Society and Natural Resources, 29*(10), 1214–1229. https://doi.org/10.1080/08941920.2015.1107788

Brand, F. S., & Jax, K. (2007). Focusing the meaning (s) of resilience: Resilience as a descriptive concept and a boundary object. *Ecology and Society, 12*(1) https://www.jstor.org/stable/26267855

Bruun, O., & Rubin, O. (2023). Authoritarian environmentalism—Captured collaboration in Vietnamese water management. *Environmental Management, 71*(3), 538–550. https://doi.org/10.1007/s00267-022-01650-7

Burnside, W. R., Pulver, S., Fiorella, K. J., Avolio, M. L., Alexander, S. M. (Eds.) 2022. 32—C.S. Holling (1973) from Part IV—Socio-environmental research in ecology. Resilience and stability of ecological systems. *Legacy readings with commentaries* (pp. 460-482). Cambridge University Press. https://doi.org/10.1017/9781009177856.038

Castro, J. E. (2007). Water governance in the twentieth-first century. *Ambiente & Sociedade, 10*(2), 97–118. https://doi.org/10.1590/S1414-753X2007000200007

Cosens, B. A., & Williams, M. K. (2012). Resilience and water governance: Adaptive governance in the Columbia River basin. *Ecology and Society, 17*(4). https://doi.org/10.5751/ES-04986-170403

Curtin, C. G., & Parker, J. P. (2014). Foundations of resilience thinking. *Conservation Biology, 28*(4), 912–923. https://doi.org/10.1111/cobi.12321

Di Vaio, A., Trujillo, L., D'Amore, G., & Palladino, R. (2021). Water governance models for meeting sustainable development goals: A structured literature review. *Utilities Policy, 72*, Article 101255. https://doi.org/10.1016/j.jup.2021.101255

Eckerberg, K., Bjärstig, T., & Zachrisson, A. (2015). Incentives for collaborative governance: Top-down and bottom-up initiatives in the Swedish mountain region. *Mountain Research and Development, 35*(3), 289–298. https://www.jstor.org/stable/mounresedeve.35.3.289

Emerson, K., & Nabatchi, T. (2015). Evaluating the productivity of collaborative governance regimes: A performance matrix. *Public Performance and Management Review, 38*(4), 717–747. https://doi.org/10.1080/15309576.2015.1031016

Fainstein, S. S., & Lubinsky, A. (2020). The relationship between citizen participation and the just city: Can more participation produce more equitable outcomes? In M. Lauria & C. S. Slotterback (Eds.), *Learning from Arnstein's ladder* (pp. 129–147). Routledge.

Falkenmark, M., Wang-Erlandsson, L., & Rockström, J. (2019). Understanding of water resilience in the Anthropocene. *Journal of Hydrology X, 2*, Article 100009. https://doi.org/10.1016/j.hydroa.2018.100009

Fallon, A., Jones, R. W., & Keskinen, M. (2022). Bringing resilience-thinking into water governance: Two illustrative case studies from South Africa and Cambodia. *Global Environmental Change, 75*, Article 102542. https://doi.org/10.1016/j.gloenvcha.2022.102542

Fiksel, J. (2006). Sustainability and resilience: Toward a systems approach. *Sustainability: Science, Practice and Policy, 2*(2), 14–21. https://doi.org/10.1080/15487733.2006.11907980

Grassini, L. (2019). Participatory water governance between theories and practices: learning from a community-based initiative in India. *International Journal of Water Resources Development, 35*(3), 404–429. https://doi.org/10.1080/07900627.2017.1354761
Guidone A. (2019). Whose knowledge, whose water, whose life: A case study of water governance and mining in Sekhukhune, South Africa. (Master's thesis, Queen's University, Kingston, Ontario, Canada).
Harrington, C. (2017). The political ontology of collaborative water governance. *Water International, 42*(3), 254–270. https://doi.org/10.1080/02508060.2017.1309507
Harris, L. M., & Morinville, C. (2013). *Improving participatory water governance in Accra, Ghana*. Africa Initiative Policy Brief No.7. https://www.cigionline.org/static/documents/ai_pb_7.pdf
Hegazi, F. F. (2019). Authoritarian governance and the provision of public goods: Water and wastewater services in Egypt (Doctoral dissertation, Duke University, North Carolina, United States). https://hdl.handle.net/10161/1
Hill, M. (2012). *Climate change and water governance: Adaptive capacity in Chile and Switzerland* (Vol. 54). Springer.
Holling, C. S. (1973). Resilience and stability of ecological systems. *Annual Review of Ecology and Systematics, 4*(1973), 1–23. https://www.jstor.org/stable/2096802
Jiménez, S. P., Giné, R., Avello, P., Leten, J., Lymer, B. L., Schneider, K., & Ward, R. (2020). *Water, 12*(3), Article 827. https://doi.org/10.3390/w12030827
Kabi, P. (2022). How South Africa's water needs plunged Lesotho into food insecurity. *Centre for Investigative Journalism*. 1 August. https://lescij.org/2022/08/01/how-south-africas-water-needs-plunged-lesotho-into-food-insecurity%EF%BF%BC/. Accessed 28 December 2024.
Lautze, J., De Silva, S., Giordano, M., & Sanford, L. (2011). Putting the cart before the horse: Water governance and IWRM. *Natural Resources Forum, 35*(1), 1–8. https://doi.org/10.1111/j.1477-8947.2010.01339.x
Lo, K. (2015). How authoritarian is the environmental governance of China? *Environmental Science & Policy, 54*, 152–159. https://doi.org/10.1016/j.envsci.2015.06.001
Lombard, A. (2024). Who benefits from Lesotho's 'white gold'? *BBC Focus on Africa*, 22 August. https://www.bbc.com/news/articles/c51yl79m0gzo
Maher, R., & Buhmann, K. (2019). Meaningful stakeholder engagement: Bottom-up initiatives within global governance frameworks. *Geoforum, 107*, 231–234. https://doi.org/10.1016/j.geoforum.2019.06.013
Mokhethi, S., & Kabi, P. (2021). Water, water everywhere… but not a drop to drink. Disease stalks villages, as Lesotho sells water to South Africa. *Center for Collaborative Investigative Journalism*. 31 March. https://ccij.io/article/water-water-everywhere-but-not-a-drop-to-drink/
Mokoena, A. M. (2022). The geography of inequality in Cape Town: A case study of access to water in Khayelitsha. (Master's thesis, University of Cape Town)
Morinville, C., & Harris, L. M. (2014). Participation, politics, and panaceas: Exploring the possibilities and limits of participatory urban water governance in Accra, Ghana. *Ecology and Society, 19*(3) https://www.jstor.org/stable/26269615
Mphambukeli, T. N. (2019). Social justice in planning: Access to adequate drinking water and contested belonging in emerging communities of Mangaung in post-apartheid South Africa. *African Renaissance, 16*(Special Issue 2) https://hdl.handle.net/10520/EJC-1a0a6686b0, 51–69.
Organisation for Economic Co-operation and Development. (2015). *OECD principles on water governance*. https://www.oecd.org/en/topics/governance.html
Özerol, G., Vinke-de Kruijf, J., Brisbois, M. C., Flores, C. C., Deekshit, P., Girard, C., Knieper, C., Mirnezami, S. J., Ortega-Reig, M., Ranjan, P., Schröder, N. J. S., & Schröter, B. (2018). Comparative studies of water governance. *Ecology and Society, 23*(4) https://www.jstor.org/stable/26796888
Pahl-Wostl, C. (2006). The importance of social learning in restoring the multifunctionality of rivers and floodplains. *Ecology and Society, 11*(1) https://www.jstor.org/stable/26267781

Pahl-Wostl, C., Craps, M., Dewulf, A., Mostert, E., Tabara, D., & Taillieu, T. (2007). Social learning and water resources management. *Ecology and Society, 12*(2) https://www.jstor.org/stable/26267868

Rana Md, M. M. P., & Piracha, A. (2020). Supplying water to the urban poor: Government's roles and challenges of participatory water governance. *Cities, 106*, Article 102881. https://doi.org/10.1016/j.cities.2020.102881

Rijke, J., Farrelly, M., Brown, R., & Zevenbergen, C. (2013). Configuring transformative governance to enhance resilient urban water systems. *Environmental Science & Policy, 25*, 62–72. https://doi.org/10.1016/j.envsci.2012.09.012

Rodina, L. (2019). Defining "water resilience": Debates, concepts, approaches, and gaps. *WIREs. Water, 6*(2), Article e1334. https://doi.org/10.1002/wat2.1334

Rockström, J., Falkenmark, M., Folke, C., Lannerstad, M., Barron, J., Enfors, E., Gordon, L., Heinke, J., Hoff, H., & Pahl-Wostl, C. (2014). *Water resilience for human prosperity*. Cambridge University Press.

Rogers, P., & Hall, A. W. (2003). Effective water governance. *Global Water Partnership Technical Committee (TEC)*. https://www.gwp.org/globalassets/global/toolbox/publications/background-papers/07-effective-water-governance-2003-english.pdf

Saikia, P., & Jiménez, A. (2023). Governance attributes for building water resilience: A literature review. *Water International, 48*(7), 809–838. https://doi.org/10.1080/02508060.2023.2274162

Salinas Rodriguez, C. N. A., Ashley, R., Gersonius, B., Rijke, J., Pathirana, A., & Zevenbergen, C. (2014). Incorporation and application of resilience in the context of water-sensitive urban design: Linking European and Australian perspectives. *WIREs Water, 1*(2), 173–186. https://doi.org/10.1002/wat2.1017

Sarami-Foroushani, T., Balali, H., Movahedi, R., Kurban, A., Värnik, R., Stamenkovska, I. J., & Azadi, H. (2023). Importance of good groundwater in economic development: The case of western Iran. *Groundwater for Sustainable Development, 21*, Article 100892. https://doi.org/10.1016/j.gsd.2022.100892

Schneider, F., Bonriposi, M., Graefe, O., Herweg, K., Homewood, C., Huss, M., Kauzlaric, M., Liniger, H., Rey, E., Reynard, E., Rist, S., Schadler, B., & Weingartner, R. (2015). Assessing the sustainability of water governance systems: The sustainability wheel. *Journal of Environmental Planning and Management, 58*(9), 1577–1600. https://doi.org/10.1080/09640568.2014.938804

Schulz, C. (2019). Governance-related values as dimensions of good water governance. *WIREs. Water, 6*(1), Article e1322. https://doi.org/10.1002/wat2.1322

Stockholm International Water Institute (SIWI). (2016). *Water governance*. UNDP-SIWI Water Governance Facility. https://siwi.org/wp-content/uploads/2016/08/issue-sheet-water-governance-web-1.pdf

Stockholm International Water Institute (SIWI). (2024). *The four dimensions*. UNDP-SIWI Water Governance Facility. https://siwi.org/undp-siwi-water-governance-facility/what-is-water-governance/the-four-dimensions

The White House. (2011). Presidential policy directive (PPD)-8: *National preparedness*. US Department of Homeland Security. https://www.dhs.gov/xlibrary/assets/presidential-policy-directive-8-national-preparedness.pdf

Tyhotyholo, T., & Ncube, B. (2023). The rhetoric of community participation in urban South African water governance. *Utilities Policy, 82*, Article 101573. https://doi.org/10.1016/j.jup.2023.101573

Van Buuren, A., Van Meerkerk, I., & Tortajada, C. (2019). Understanding emergent participation practices in water governance. *International Journal of Water Resources Development, 35*(3), 367–382. https://doi.org/10.1080/07900627.2019.1585764

Walker, J., & Cooper, M. (2018). Resilience. In N. Castree, M. Hulme, & J. D. Proctor (Eds.), *Companion to environmental studies* (pp. 90–94). Routledge.

Wiek, A., & Larson, K. L. (2012). Water, people, and sustainability—a systems framework for analyzing and assessing water governance regimes. *Water Resources Management, 26*, 3153–3171.

Woodhouse, P., & Muller, M. (2017). Water governance—An historical perspective on current debates. *World Development, 92*, 225–241. https://doi.org/10.1016/j.worlddev.2016.11.014

Yan, J., Li, R., & Ran, R. (2022). When collaborative water governance meets authoritarian environmentalism: The dilemma of safe water supply project in coal mining villages of China's Shanxi Province. *Sustainability, 14*(3), Article 1277. https://doi.org/10.3390/su14031277

Zwarteveen, M., Kemerink-Seyoum, J. S., Kooy, M., Evers, J., Guerrero, T. A., Batubara, B., Biza, A., Boakye-Ansah, A., Faber, S., Flamini, A. C., Cuadrado-Quesada, G., Fantini, E., Gupta, J., Hasan, S., Ter Horst, R., Jamali, H., Jaspers, F., Obani, P., Schwartz, K., Shubber, Z., Smit, H., Torio, P., Tutusaus, M., & Wesselink, A. (2017). Engaging with the politics of water governance. *Wiley Interdisciplinary Reviews Water, 4*(6), 1–9. https://doi.org/10.1002/wat2.1245

Chapter 3
Urban Legislation and Policy Framework Guiding Water Governance in Postcolonial Zimbabwe

Abstract This chapter examines the legislation and policy framework guiding urban water governance in postcolonial Zimbabwe. The analysis is enabled by a desktop review of key policies and acts of parliament that regulate and guide the governance of the urban water system in Zimbabwe. The ideologies informing the respective legislation and policies are examined to understand the motivation and expected outcomes, focusing on citizens' rights to adequate water and what it means. The chapter is framed in the context of distributive justice and articulates how the policies and legislation relate to human needs and social justice. The findings from the analysis show that the policies address multiple issues related to water governance in urban areas, including poor implementation of good policies, lack of transparency and accountability associated with water governance, and politicization of water governance, much to the detriment of effective water distribution. The continued neglect of the water sector greatly compromises water delivery and shows the distributional inequalities and injustices imposed on people living in poverty who fail to access quality and potable water.

Keywords Urban policy · Water policies · Water governance · Politics of difference

3.1 Introduction

Zimbabwe is a former British colony. When the country gained its independence in 1980, the government inherited the British system of urban governance, which was enacted by legislation. The chapter traces the legislation and policy framework guiding urban water governance in Zimbabwe. An analysis of the legislation and policy frameworks is critical in urban water governance. Schultz (2019) has argued that understanding water governance is effectively done by describing a combination of institutional frameworks (to be discussed in Chap. 4), water politics (discussed in subsequent Chaps. 5 and 7) and water policy (the focus of this chapter). Thus, this chapter analyses the historical context of water governance in Zimbabwe, focusing

A. R. Matamanda et al., *Urban Water Governance in Postcolonial Zimbabwe*, Local and Urban Governance, https://doi.org/10.1007/978-3-031-96973-7_3

on the colonial and postcolonial historical events that have shaped the existing water governance policies in the country. Next, the chapter explores the different mechanisms and instruments used to achieve certain outcomes. This includes international conventions such as the Sustainable Development Goals (SDGs), to which Zimbabwe is a signatory, and is often applied in the development of several national policies, including water policies. While water policies guide the planning, use, and management of water resources, their formulation tends to be informed by different factors, often with intended outcomes that may not always be achieved due to multiple factors such as politics of difference, lack of funding, and capacity to implement the policies. Occasionally, external factors such as international conventions may require a shift and realignment of the policy objectives.

Therefore, this chapter traces the historical development of urban water policy in postcolonial Zimbabwe. The respective policies are analyzed, starting with the international and regional policies that influence water policies at the national level. These international policies are critical as they set the international standards and guidelines critical for benchmarking local practices. However, some regulations and goals of these international policies may not always be applicable in certain contexts; hence, the need to localize them as is often the case with the SDGs, of which some are considered as being difficult to achieve in most African countries, including Zimbabwe. Lastly, the chapter analyses the national policies in Zimbabwe, looking at the policies and acts of parliament that regulate the planning and management of the urban water sector in Zimbabwe. This analysis is guided by the theory of distributive justice, which focuses on how resources (water) are distributed in society. This distribution is determined by power dynamics in which certain individuals and groups control the allocation of resources. The chapter ends with a discussion of the emerging issues and the policy recommendations for a sustained and resilient water policy framework for Zimbabwe.

3.2 Social Justice in Water Governance

Water policies relate to sociospatial justices as the policymakers constitute a group of people who must determine the regulation, distribution and management of water resources with their intended outcomes. The formulation of the policies is often undertaken by the privileged elites, whose ideology and aspirations are not always in the interest of the public. Soja (2009) argued that injustices are inherent in how locational discrimination plays out where "biases are imposed on certain populations because of their geographical location". This perspective of injustice is evident in how certain locations are disadvantaged in terms of water access; for example, most informal settlements in the Zimbabwean context are not serviced with reticulated water systems (Maruta et al., 2023; Ndoziya et al., 2019). On the other hand, due to the long-standing marginalization of Matebeleland, successful supply has been attributed to hydropolitics (Musemwa, 2008). This illustrates why the Zimbabwe African National Union–Patriotic Front (ZANU–PF) government has not made much effort to provide access to potable water in Bulawayo, a situation that has persisted for many years. The locational discrimination is also noted in how the city of Harare disproportionately imposes water rationing across the city with low-income suburbs such as Mabvuku, Tafara and Budiriro experiencing longer spells of water cuts than the northern suburbs such as Borrowdale and Mt. Pleasant (Human Rights Watch, 2013).

Second, social justice is explained by Soja (2009) as a product of the political organization of space in which institutionalized residential segregation is evident, redlining urban investments and the imprint of colonial social control. Regarding the water sector, such politicization of space and resources results from how politics take precedence in the policymaking process, policy implementation, and ultimately, by disadvantaging individuals and groups that do not support or sympathize with the ruling party. As explained earlier, the case of Bulawayo is a typical example where politics of difference have had an impact on the water sector in the city since the early 1980s.

Third, Marcuse (2010) explained how the unequal distribution of resources based on race, gender, religion, or political affiliation results in the unequal treatment of residents and the forced limitations placed on certain individuals and groups to explore life possibilities. In this case, access to resources is planned, managed and distributed in a segregated manner based on various criteria. Water pricing can also be an issue in which some communities are levied more than others while not receiving water services.

Lastly, distributive justice is associated with capitalist urbanization in which cities are considered profit entities in which there is always surplus capital and profit to be generated (Soja, 2010). In this regard, resources in cities tend to be commercialized and enjoyed by those individuals who can afford to pay more to access the goods and services.

3.3 Historical Context of Water Governance in Zimbabwe

3.3.1 Precolonial Legacy

Before the arrival of European colonizers, Zimbabwe's Indigenous societies managed water using community-based systems that were deeply embedded within the social and cultural fabrics of the tribes (Musemwa, 2008). These systems were primarily governed by traditional leaders and relied on well-established customary laws and practices. Water sources like rivers, springs, and rain were considered sacred, often associated with spirits and ancestors, which influenced how communities respected and managed these resources. The Shona and Ndebele people, for instance, practiced rainmaking ceremonies that symbolically and practically addressed water management, highlighting the integrative approach of spirituality and resource stewardship (Muyambo & Maposa, 2014). Traditional management systems were adapted to the region's semi-arid climate, focusing on water conservation techniques such as dam and weir construction using locally available materials. These structures were essential for irrigation, especially in the agriculturally rich regions, and were maintained communally, illustrating a collective responsibility towards water resource management. The systems exemplified sustainable practices as they balanced consumption with conservation, ensuring that water was available for all community members.

3.3.2 *Colonial Legacy*

The colonial era introduced a profound shift in the management of water resources in Zimbabwe. With the establishment of the British South Africa Company in the late nineteenth century, water governance began to mirror the administrative structures of the colonial government, which prioritized commercial agriculture and mining operations (Worster & Empire, 1985). The introduction of cash crops like tobacco and maize necessitated large-scale irrigation schemes, which were often implemented at the expense of local communities' access to water (Chamboko et al., 1999). Colonial policies enforced a centralized water management system, stripping traditional leaders of their roles in water governance and reallocating control to colonial administrators. This was characterized by the enactment of the Water Act of 1927, which established the legal foundation for water rights that heavily favored white settlers and commercial farmers (Musemwa, 2008). Under this system, water rights were tied to land ownership, which disenfranchised most of the Indigenous population who did not own land. Furthermore, the colonial administration often constructed large dams and reservoirs without considering the social and environmental impacts on local communities, leading to displacement and alterations in the local ecosystems (Mashingaidze, 2019).

The colonial approach to water management in Zimbabwe not only disregarded traditional knowledge systems but also laid the groundwork for inequalities that persisted long after independence. The legacies of these policies continue to influence the frameworks and challenges of postindependence water governance, where issues of access, equity, and sustainability remain critical. The transition from precolonial to colonial water management practices clearly shows that the historical context significantly shapes the current challenges in Zimbabwe's urban water governance. Understanding these historical dynamics is crucial for effectively addressing contemporary issues, as they offer insights into the sociocultural and political complexities that any effective water governance framework must navigate.

3.3.3 Postindependence Water Policies

Following Zimbabwe's independence in 1980, the new government faced the daunting task of restructuring the country's water governance system to redress the inequalities established during the colonial era (Musemwa, 2008). The initial postindependence period was marked by ambitious efforts to increase water access to underserved and rural populations, which involved significant changes to the policy and governance frameworks. One of the first major legislative changes was the amendment of the Water Act in 1981, which aimed to decentralize water management and increase community participation in water governance. This policy shift was intended to democratize access to water resources, reflecting the government's commitment to social equity and justice. The government also embarked on a rural water supply and sanitation program, which saw the drilling of boreholes and construction of small dams to enhance water availability in rural areas.

The introduction of the Economic Structural Adjustment Program in the 1990s brought a neoliberal shift to water governance, emphasizing cost recovery and the privatization of water services (Bhanye et al., 2023). This period saw the introduction of water tariffs and the promotion of private sector participation in water supply. While these policies were aimed at making water management more efficient and financially sustainable, they also led to increased water prices, making access more difficult for poor urban and rural populations (Chikozho & Latham, 2005). In 1998, the ZINWA was established to oversee all water management activities in the country. ZINWA was tasked with managing both urban and rural water systems, with a mandate that included water allocation, quality monitoring, and infrastructure development. The creation of ZINWA was part of a broader effort to centralize water management and improve the coordination of water resources at the national level.

The adoption of the new Constitution of Zimbabwe (COZ) in 2013 marked a significant milestone in Zimbabwe's water governance, as it explicitly recognized access to clean, safe, and potable water as a basic human right. The COZ obligates the state to take reasonable legislative and other measures, within its resources, to achieve the progressive realization of this right. This legal acknowledgement has been pivotal in shaping subsequent water policies and has provided a framework for advocacy and legal action when the government fails to meet its water provisioning obligations. Despite these reforms, Zimbabwe's water governance system faces ongoing challenges, including inadequate infrastructure, corruption, and a lack of funding. Urban areas, in particular, experience frequent water shortages and deteriorating water quality due to ageing infrastructure and insufficient investment in maintenance and upgrades. The challenge is exacerbated by periodic droughts, which stress the already strained water management systems.

3.3.4 Impact of Economic and Political Factors

Zimbabwe's economic challenges and political instability have had profound effects on water governance, complicating the management and distribution of water resources. The relationship between economic policies, political decisions, and water governance shows the broader issues of resource management in contexts of instability and economic strain. The severe economic decline that began in the early 2000s, marked by hyperinflation, significant drops in industrial and agricultural production, and a consequential lack of foreign investment, had a direct impact on the country's ability to maintain and upgrade water infrastructure (Bhanye et al., 2023). As economic conditions worsened, government funding for critical infrastructure, including water treatment facilities and piping systems, dwindled. This led to the widespread deterioration of the water supply systems, especially in urban areas where the demand was highest. Frequent breakdowns, leaks, and the inability to treat water adequately became more common, leading to intermittent water supplies and increased waterborne diseases among the population (Matsa et al., 2021).

Political instability, characterized by frequent changes in leadership within key ministries and shifts in policy direction, further exacerbated water governance issues. Political interference often influenced the management positions within ZINWA and the local water boards, leading to a politicization of water management (Hadebe et al., 2015; Muchadenyika & Williams, 2018). This instability undermined technical decision-making and continuity in policy implementation, as new administrations often reversed or altered the policies of their predecessors without a clear rationale other than political expediency (Mapedza et al., 2016). Corruption is another significant factor affecting water governance. Instances of misallocated budgets, bribery in the awarding of water-related contracts, and embezzlement of funds meant for water infrastructure projects have been reported. For example, in 2021 ZINWA was involved in a tender scandal after controversially awarding a US$109 million tender to China Nanchang for the construction of the Kunzvi Dam when there was a lower bid of US$66 million offered by another company (*ZimEye*, 2021). Such practices not only drain financial resources but also undermine public trust in institutions responsible for water governance. The lack of transparency and accountability in the management of water resources discourages investment and hinders effective regulation and maintenance of water services (Mapedza et al., 2016).

The fast-track land reform program initiated in the early 2000s also had an indirect effect on water governance. The redistribution of agricultural land, while politically popular, led to disruptions in established water management systems on commercial farms, which had been integral to local water management schemes. The new landowners often lacked the resources or expertise to maintain these systems, leading to decreased water efficiency and increased conflicts over water use in rural and peri-urban areas (Manzungu & Dzingirai, 2012).

In response to these ongoing challenges, the government and international partners have occasionally launched initiatives aimed at improving water access and infrastructure, especially during crises of acute water shortages or outbreaks of waterborne diseases like cholera. However, these efforts have often been reactive rather than proactive, addressing symptoms rather than the systemic issues at the root of water governance problems (Bhanye et al., 2023).

3.4 International Conventions Guiding Urban Water Governance

3.4.1 Sustainable Development Goals

The SDGs are a set of goals enacted by world leaders in 2015 to advance global development, with the main object of 'leaving no one behind'. Many governments, including the Government of Zimbabwe, have adopted the SDGs as a development frame to guide socioeconomic development and provide benchmarks for development milestones (Government of Zimbabwe, 2021). The SDGs are a broader framework guiding water governance in Zimbabwe. The SDGs are encapsulated and packaged into seventeen broad goals that focus on different aspects (see Table 3.1).

According to Zimbabwe's Second Voluntary National Review of the SDGs of 2021, the Government of Zimbabwe has committed to implementing 10 SDGs (2, 3, 4, 5, 6, 7, 8, 9, 13 and 17) closely monitored and prioritized in national development (Government of Zimbabwe, 2021). Specifically, SDGs 5, 6, 9 and 16 are relevant to urban water governance issues because they focus on water issues and governance. SDG 6 is often called the water SDG as it commits to ensuring the availability of sustainable water and sanitation management by 2030 (Di Vaio et al., 2021). Target 6.1 reinstates that by 2030, universal and equitable access to safe and affordable drinking water should be available for all. In this target, the focus is on ensuring that all citizens have access to potable water, which is related to the concept of distributive justice, which emphasises that the allocation of resources should be fair and just, with all stakeholders benefiting equitably without leaving anyone behind. In this regard, SDG Target 6.2 upholds the gender perspective in water use and allocation, which is a critical pillar in urban water governance considering how women tend to be left behind in decision-making processes regarding urban water planning and management; yet they are the main users who bear the brunt of accessing the water (Matamanda et al., 2024). SDG 5, Target 5.5, also emphasises the significance of gender inclusion by ensuring that women and girls have full and effective participation in decision-making and control, which includes water use and management. Hence, the planning and management of urban water resources and services should recognize the needs of all genders and not be discriminatory in the

Table 3.1 The seventeen SDGs

Goal 1	No poverty	End poverty in all its forms everywhere
Goal 2	Zero hunger	End hunger, achieve food security and improved nutrition and promote sustainable agriculture
Goal 3	Good health and well-being	Ensure healthy lives and promote well-being for all at all ages
Goal 4	Quality education	Ensure inclusive and equitable quality education and promote lifelong learning opportunities for all
Goal 5	Gender equality	Achieve gender equality and empower all women and girls
Goal 6	Clean water and sanitation	Ensure availability and sustainable management of water and sanitation for all
Goal 7	Affordable and clean energy	Ensure access to affordable, reliable, sustainable and modern energy for all
Goal 8	Decent work and economic growth	Promote sustained, inclusive and sustainable economic growth, full and productive employment and decent work for all
Goal 9	Industry, innovation and infrastructure	Build resilient infrastructure, promote inclusive and sustainable industrialization and foster innovation
Goal 10	Reduced inequalities	Reduce inequality within and among countries
Goal 11	Sustainable cities and communities	Make cities and human settlements inclusive, safe, resilient and sustainable
Goal 12	Responsible consumption and production	Ensure sustainable consumption and production patterns
Goal 13	Climate action	Take urgent action to combat climate change and its impacts
Goal 14	Life below water	Conserve and sustainably use the oceans, seas and marine resources for sustainable development
Goal 15	Life on land	Protect, restore and promote sustainable use of terrestrial ecosystems, sustainably manage forests, combat desertification, and halt and reverse land degradation and halt biodiversity loss
Goal 16	Peace, justice and strong institutions	Promote peaceful and inclusive societies for sustainable development, provide access to justice for all and build effective, accountable and inclusive institutions at all levels
Goal 17	Partnerships for the goals	Strengthen the means of implementation and revitalize the global partnership for sustainable development

Source: United Nations (2015)

policy outcomes and planning process, which should be participatory and inclusive (Adams et al., 2018).

Target 6.5 brings to attention the issues around integrated water resources management (IWRM), which should be undertaken at all levels and directly speaks to the values and principles of good governance as it promotes inclusivity and participatory processes. Lastly, urban governance issues are espoused in SDG 16, focusing on peaceful and inclusive societies for sustainable development while also providing access to justice for all and the building of effective, accountable and inclusive institutions at all levels. SDG 16 directly speaks about the emerging issues in governance as it raises critical issues on the commitment to reducing corruption and bribery, which is a significant challenge. For example, corruption and embezzlement of public funds tend to have negative implications on service delivery and water governance as has been evidenced in the case of Johannesburg in South Africa and Harare in Zimbabwe where corruption and misuse of public funds have greatly compromised water access over the past years (Masimba Nyama & Walnycki, 2024; Pieterse, 2019). SDG 16 recognizes the need for developing effective, accountable and transparent institutions at all levels. This is a critical pillar of good governance since institutions are responsible for formulating and coordinating policies that guide and regulate the water sector. Target 16b promotes enforcement of non-discriminatory laws and policies for sustainable development.

The government of Zimbabwe has localized the local policies to align with the SDGs, a situation that affirms the commitment of the country to ensure that development is informed by the broader goals and targets set in the SDGs, as explained above. These include development frameworks, including the Zimbabwe Agenda for Sustainable Socio-Economic Transformation of 2013–2018, the National Development Strategy 1 of 2021–2025, and the national budgets (see Government of Zimbabwe, 2020), which aspires to transform Zimbabwe into a middle-income country by 2030 and good governance and social justice are upheld as critical pillars in achieving this growth which aligns with the ideals and values of the SDGs as discussed in the foregoing paragraph. It is also crucial to note that the City of Harare has committed to monitoring the city's development and benchmark it with the aspirations of the SDGs. This is evident from the *Harare Voluntary Local Review of Sustainable Development Goals (SDGs) Report of 2020*, which analyzed the city's progress considering the SDGs (City of Harare, 2020). Water governance was among the issues discussed in the report, showing the relevance of the SDGs in guiding urban governance matters in the city.

3.4.2 New Urban Agenda of 2016

The New Urban Agenda (NUA) is a key document applied to guide global urban development. NUA is a document that was compiled in October 2016 at the Habitat III Cities Conference in Quito, Ecuador (see United Nations, 2017). The rapid urbanization currently experienced in most developing countries is recognized in the NUA and the myriad challenges posed by this rapid urbanization and the need for innovative urban planning, management, and governance practices that address these challenges. In this regard, the NUA envisages to "achieve cities and human settlements where all persons are able to enjoy equal rights and opportunities" (United Nations, 2017:5). This aspiration recognizes the need for upholding human rights and maintaining equity in which all citizens have equal access to opportunities and quality of life.

Urban governance is mentioned in several places in the NUA, and urban water governance falls under this category of urban governance. Several principles of urban water governance are identified in the NUA. First, section 15(c) commits to adopting sustainable, people-centered, age- and gender-responsive and integrated approaches to urban and territorial development, planning and governance (United Nations, 2017). Here, the NUA recognizes the significance of strengthening urban governance through sound institutions, inclusive approaches, mechanisms that empower citizens and predictable systems. In essence, these principles are critical for water governance as they ensure that institutions are capacitated and can administer the water issues and have systems that recognize the value and contributions of different stakeholders and are thus impartial and apolitical. Second, section 85 considers the perspectives of decentralization and multilevel urban governance systems, which is crucial for sustaining advancing urban governance due to the clearly defined systems that also enhance accountability and transparency. Third, the NUA upholds the value of evidence-based governance, which is significant because, in many instances, especially in the Global South, policies tend to be premised on a political basis with little evidence supporting the rationality of the decisions. Lastly, section 160 of the NUA touches on a critical issue in contemporary urban governance with much relevance to urban water governance. Technology is a critical issue that helps advance urban water governance systems. This can be done through e-participation in water resource allocation, water pricing, or billing. To some extent, this principle has been applied in Zimbabwe but without much success. For example, the use of prepaid water meters has been met with mixed responses in Zimbabwean cities (Maramura, 2018; Reniko & Kolawole, 2020). On the other hand, several municipalities, including the City of Harare, are increasingly adopting the e-billing and payment system, though with some challenges (Matamanda, 2022).

3.4.3 African Agenda 2063

The African Agenda 2063 is a strategic framework commissioned by African leaders in 2013 to advance socioeconomic development in Africa with aspirations set for 2063. The agenda is based on seven aspirations that guide development and emphasize the creation of a prosperous Africa, the core of which should be based on inclusive growth and sustainable development (DeGhetto et al., 2016). Aspiration 3 focuses on creating "An Africa of good governance, respect for human rights, justice and the rule of law" (African Union Commission, 2015:14). Specifically, the dimensions of good governance are to be achieved through consolidation of democratic gains and improvement of the quality of governance, respecting human rights and the rule of law. Royo et al. (2022) have highlighted the linkages between the SDGs and the African Agenda 2063, where the advancement of good governance is a common theme, while the drive towards strengthening institutions and promoting good governance is critical for sustained development. Moreover, the Africa Agenda 2063 also aspires and recognizes the significance of building development-oriented institutions that help advance good practices and uphold human values and well-being (African Union Commission, 2015).

3.5 National Policies and Legislation Regulating Water Governance

3.5.1 Constitution of Zimbabwe Amendment No. 20 of 2013

Amendment No. 20 of the COZ is the supreme law governing and regulating all affairs in Zimbabwe, including urban water governance. The previous constitution has been regarded as perpetuating injustices because it resulted from a compromise reached at the Lancaster House Conference in 1979 and lacked meaningful public input. Thus, the public participation engaged in the making of the COZ has been hailed as critical in advancing justice through integrating the needs, concerns and aspirations of different stakeholders and groups of citizens in Zimbabwe. This process was undertaken over 4 years (2009–2012), showing the robustness of the participation process, which was steered by the Government of National Unity and constituted three major political parties in Zimbabwe at the time (Constitution Select Committee, 2012). These political parties included ZANU–PF, the Movement for Democratic Change (MDC) and the MDC–Tsvangirai, illustrating the political inclusivity and diversity guiding the COZ. In this regard, it is evident that the COZ was people-centered and sought to uphold social justice through inclusivity and the ability to integrate the concerns of all parties involved, thereby addressing issues of discontent and unequal treatment as articulated by Marcuse (2010).

However, some proponents (Shonhe, 2011) have argued that despite the consultation and participatory process, the COZ has remained partisan and has not fully integrated the citizens' concerns. Nyabeze (2015:19) pointed out that the "major setback to the realization of the already mentioned gains if not cured in time will be the re-alignment of the subsidiary legislation with the new norms and basic values stated in the COZ." In this regard, while the COZ provides new laws and provisions, little attempt has been made to revise the other legislation, which makes it difficult to implement effective reforms in most circumstances.

Section 2(2) of the COZ stipulates that its obligations are binding upon every individual, organization and institution, including the State, which has to play its role. The COZ recognized a range of issues and statutory obligations which relate to the planning of human settlements together with the provision of basic services such as water. The State is thus empowered to oversee the regulation and management of urban water systems as has been evident from the 2013 situation where the Minister of Local Government, Dr. Ignatius Chombo, acting on behalf of the State, called upon all local authorities to scrap the debts owed by citizens after realizing that these amounts were not going to be recovered (see Maponga, 2013). Minister Chombo's decision to scrap approximately US$2 billion owed by citizens to the 92 local authorities across the country was done in terms of the Prescription Act (Chapter 13). While this was identified as a political move, for example, the City of Harare scrapped US$350 million owed by rates and taxpayers (Kadungure, 2013; Muchadenyika & Williams, 2018), although there were some unjustified increases and debt accumulation accrued by residents during the transition from the

Zimbabwean dollar to the US$ in 2009, much to the disadvantage of the citizens. These justifications were also attributed to the Declaration of Human Rights in Section 77 (a) of the COZ, which recognizes the socioeconomic rights to water and states:

> Every person has a right to safe, clean and potable water and the State must take reasonable legislative and other measures, within the limits of the resources available to it, to achieve the progressive realization of this right.

In this instance, the Minister sought to address issues around capitalist urbanization (Soja, 2010) in which residents had been unfairly charged by the urban councils when their accounts transitioned from Zimbabwean dollars to US$ with many estimations being made, which subsequently disadvantaged the residents.

The COZ has also emphasized principles of good governance that are evident in section 194 in which the basic values and principles governing public administration are stipulated. Specifically, section 194(1) states that public administration in all tiers of government, including institutions and agencies of the state and government-controlled entities and other public enterprises, must be governed by the democratic values that include the principles presented in Table 3.2.

According to section 194(1) (g) of the COZ, cooperation and coordinated efforts should be upheld in the governance process, including urban water governance such that different institutions, actors and stakeholders work together towards a common goal. This has been evident in instances where local governments have worked with NGOs, community members and other stakeholders such as academics in trying to address the water challenges in the country. However, the situation has at times tended to be compromised by politics of difference where the central government has often frustrated efforts and initiatives by the local governments, especially those led by the opposition parties depicting a lack of shared interest and the ability to work together (Muchadenyika & Williams, 2018). The following quote by Musemwa (2021:35–36) shows the lack of cooperation and desire to collaborate and work together, which has largely undermined the success of good urban water governance in the country:

> Upon seeing signs that he was moving swiftly to repair broken infrastructure and dilapidated amenities, resurfacing roads, restoring street lights, fixing water leaks, and having meetings

Table 3.2 Principles of good governance set in section 194(1)(a–h) of the Constitution

(a)	A high standard of professional ethics must be promoted and maintained;
(b)	Efficient and economical use of resources must be promoted;
(c)	Public administration must be development-oriented;
(d)	Services must be provided impartially, fairly, equitably and without bias;
(e)	People's needs must be responded to within a reasonable time, and the public must be encouraged to participate in policymaking;
(f)	Public administration must be accountable to parliament and to the people;
(g)	Institutions and agencies of government at all levels must co-operate with each other;
(h)	Transparency must be fostered by providing the public with timely, accessible and accurate information.

with international donors to bring investment and aid to Harare, the minister of local government, Ignatius Chombo, fired him in April 2003.

Section 194(1)(d) focuses on the need to provide services to all citizens equitably and without any bias. Focusing on water supply, the government has failed dismally to provide this service as there has been some marginalization of certain groups of people and communities, which highlighted the urban water injustices that emerged from the actions taken by the government in addressing water woes. For example, Bulawayo and the Matabeleland province at large have remained largely underserviced in water provision, a situation described by Musemwa (2006) as hydropolitics, which describes the deliberate actions of the government to 'discipline' the dissident city of Bulawayo. Moreover, in Harare, the high-density suburbs are often less prioritized in water distribution while the affluent suburbs are serviced. Evidence is seen in suburbs such as Mabvuku, Tafara and Budiriro, which have experienced dry taps that have lasted for more than 6 months in which they did not receive any running water (Human Rights Watch, 2013).

3.5.2 National Water Policy

The National Water Policy of 2013 (NWP) recognizes the role of adequate water supply in urban development. The NWP was developed during the Government of National Union and recognizes the collapse of the urban water sector and the need to rehabilitate the water sector, which was once hailed as the best in southern Africa. Efforts to rehabilitate the water sector set in the NWP include the recovery phase that provides recovery activities through which the continued deterioration of the water and sanitation sector and infrastructure are supposed to be rehabilitated before they reach a point where they will be irrecoverable (Government of Zimbabwe, 2012). The recognition of the recovery phase has been critical, considering the collapse of the urban water system in most urban centers across the country, hence the need to rehabilitate this infrastructure and ensure citizens have access to potable water. A critical issue relating to the water policy has been unequal treatment and forced constraints (Soja, 2009), where most urban areas were subjected to water rationing for longer periods. The collapse of the water sector in most cities was also considered a violation of human rights and gross injustice, especially after the 2008/09 cholera outbreak, which resulted in at least 4000 deaths in Harare.

Through the recovery phase, the NWP envisages engaging in recovery activities through which the continued deterioration of the water and sanitation sector and infrastructure are rehabilitated before they reach a point where it would be difficult to recover them. Urban local authorities are therefore tasked to ensure efficient, affordable and sustainable access to water services in areas of their jurisdiction. With regard the urban housing developments, the NWP opts to maintain high standards and call for a moratorium on new housing developments that do not meet existing standards stipulated in the various national statutes and legislative guides

(Government of Zimbabwe, 2012:13). Maintenance of these high standards aligns with the values of social justice that seek to uphold the equal distribution of water resources to all households without any segregation.

Efforts to uphold the water quality are enshrined in section 7.1.4, which emphasizes issues on capacity building in which government institutions and parastatals directly dealing with water services are to be strengthened and rebuilt to enable them to deal with the water issues in the country. The government also committed to providing the financial resources to enhance capacity building in the water sector. This initiative has been set to be complemented by research and development. The NWP has called for the establishment of a water research fund to be financed with funds from a levy on water revenues accruing to ZINWA and the Environmental Management Agency.

While ZINWA had been granted responsibility for water governance, the NWP in section 7.4.1 reallocated responsibility for water services to local governments (Government of Zimbabwe, 2012). The policy emphasized that local authorities could choose to contract ZINWA as a service provider but does not compel them to do so, thereby emphasizing the central role of urban councils in managing their water supply systems. In this way, the NWP decentralized power and autonomy to the urban councils, contrary to the ZINWA Act (as will be discussed in the forthcoming sections). However, regardless of this decentralization, the central government has continued to interfere in the water affairs of most cities across the country. Taruvinga (2024) found that:

> Zimbabwe's attempts at decentralizing water governance have not been fully implemented. This is because of resistance from the ruling party, Zanu–PF, which fears losing control over water. The result is that community participation in local water systems is undermined.

This point by Taruvinga confirms the sentiments made by Tom and Munemo (2015) in their review of the NWP. They lauded the policy for containing good governance principles but argued that the implementation of the policy is problematic and contradicts the values and aspirations set out in the NWP. This instance shows how social injustice has resulted from the politicization of the water sector where the ZANU–PF ruling party has dominated the urban water sector, including revenue collection through Zimbabwe, and not investing the profits in sustaining the water infrastructure and services.

Additionally, the NWP in section 7.4.5 indicated that local authorities have to ring-fence revenue from water sales and set tariffs to achieve full cost recovery.

> *Local authorities will undertake cost accounting and will ring-fence revenue from water sales. Tariffs will be set to achieve full cost recovery in the normalized phase. The long-term financial viability of service delivery will depend largely on the financial well-being of consumers and their ability to pay in full for services they use* (Government of Zimbabwe, 2012:24).

However, the issue of cost recovery is premised on a capitalist ideology in which water is commodified and commercialized (Mukonavanhu et al., 2021). Tom and Munemo (2015) have argued that the issue of cost recovery created social segregation as certain individuals and groups are disadvantaged with the water pricing, which for long has been unfairly applied in most Zimbabwean cities. The issue of cost-recovery highlighted in the NWP is linked to the unequal distributive outcomes of capitalist urbanization (Soja, 2010) in which water has been commercialized. In this way, many households have been grappling with water bills as the urban councils have commercialized water.

While the NWP calls for cost recovery on water use, most local authorities in Zimbabwe have been billing residents, yet not delivering any water to them. For example, in Harare, the council continues to bill households monthly charges of US$50 and more for water services they do not receive, and when the water is running in the tap, it is often not suitable for drinking, leaving residents to find alternatives for water access (Gogo, 2019). On the other hand, there is a lack of transparency in the billing system and the use of revenue generated from the water levies, which undermines the aspirations of the NWP (Brooke and Fenner, 2023).

3.5.3 Water Act of 1998

The Water Act [Chapter 20:24] of 1998 is the main legislation that provides guidelines for the development and utilization of water resources in Zimbabwe. According to section 3 of the Water Act, all water in Zimbabwe is vested in the President. This section greatly empowers the President with absolute powers over water resources in the country, emphasizing the authoritarian role that the President can take regarding water resources. Moreover, according to section 4 of the Water Act, there is no private ownership of water in Zimbabwe. Individuals can be granted water permits that only allow them the right to use the water following the provisions of the permit. This section recognizes the fundamental role of water as a public common, which cannot be privately owned, such that all citizens have equal access without any segregation. The permits include those that individuals can apply to install boreholes on their properties as specified in section 35 of the Water Act, which highlights the need for individuals to apply for permission to sink, alter or deepen a borehole or any wells on their properties; thus, illustrating the significance of the catchment councils in monitoring water abstraction.

The Minister is empowered to develop policies that guide the orderly and integrated planning of utilization and protection of the country's water resources. Section 6(1)(a-b) recommends that the Minister should ensure that all citizens have access to water for primary purposes, which includes water for domestic use. In this regard, the Minister is tasked to develop IWRM plans that ensure that water is accessed by all citizens without any segregation. The Minister is also obliged to secure the provision of affordable water to consumers in underprivileged communities such as informal settlements. While several measures have been undertaken by the Minister to ensure water access, the case of Bulawayo and the greater parts of Matabeleland reveal limited commitment from the Minister to service the underresourced areas (Musemwa, 2008), while the persisting water challenges in most low-income neighbourhoods in major cities across Zimbabwe shows the failure of the state to ensure that citizens have adequate access to potable water.

Part III of the Water Act provides guidelines for the establishment of catchment councils and their functions in water use and management. In this part of the act, more powers are bestowed on the Minister, who should consult the ZINWA in the establishment of the catchment councils whose mandates include managing water resources, regulating the issuance of water use permits, monitoring the use of water resources by permit holders and ensuring that water resources are not used in ways that contravene the Water Act and eventually jeopardize the water resources. However, over the years, there seem to be conflicts in coordinating the efforts of these catchment councils. Harare is a typical example. In December 2024, water from Lake Chivero was contaminated and polluted by cyanobacteria, fueled by raw sewage discharged into the lake by the City of Harare (Zinyuke, 2024). The water pollution resulted in the death of livestock, fish and wildlife that depend on the lake for freshwater. This example shows the failure of the catchment council to regulate the discharge of the raw sewage into the lake, a problem that has persisted over the years (Magadza, 2003; Manjengwa et al., 2019). Also, these practices contradict section 67 of the Water Act, which upholds the need for environmental approaches to be considered in water resources management.

3.5.4 Zimbabwe National Water Authority Act of 1998

The ZINWA Act [Chapter 20:25] was enacted in 1998 as part of the water governance reforms in the country through the establishment of ZINWA. Specifically, the ZINWA Act was established to empower ZINWA to provide technical support regarding water resources management in the country. Section 5 of the ZINWA Act provides for the functions of ZINWA that include advising the minister of water and environment on the formulation of national policies and standards relating to water quality, borehole drilling and water resources planning. In this way, ZINWA was granted a central role relating to water resources planning and management. Additionally, section 5(e) of the ZINWA Act states that ZINWA was mandated to encourage and assist urban councils in discharging their functions regarding the development and management in their areas of jurisdiction. While this Act tasked ZINWA with these responsibilities, the realities on the ground have shown the deplorable and declining state of the urban water sector, largely owing to politicization, corruption and mismanagement showing the incompetence of ZINWA and its failure to fulfil the mandates set in the ZINWA Act. Muchadenyika and Williams (2018) argued that the nationalization of the water sector through vesting many powers in ZINWA was largely retrogressive because it was a political move meant to capture and control water infrastructure in the country. Such looting by ZINWA included the wanton removal of water pumps "from urban local authorities (for instance Ruwa) and installing them on farms owned by top ZANU–PF officials" (Muchadenyika & Williams, 2018:842).

While the ZINWA Act was intended to bring reforms in water governance, it has exacerbated the water situation in the country with increased water rationing, infrastructure collapse and cholera outbreaks associated with poor urban water governance. In their 2015 annual report, Transparency International Zimbabwe profiled the corruption cases including the 2014 Dams Scandal by ZINWA in which the parastatal was implicated for squandering US$46 million in 2014, US$49.4 million and US$4.2 million in 2007, while records show that there were 14 uncompleted dam projects nationwide (Hadebe et al., 2015). Moreover, the salary scandal also showed that top executives from ZINWA were receiving salaries of US$5500 and benefits of US$7728.

3.5.5 Public Health Act

The Public Health Act [Chapter 15:09] is an act of parliament which guides and regulates public health issues in the country. Part VI specifically outlines the provisions for water supply and water safety as well as empowering certain institutions to manage the provision of the water supply. In section 64(1), local authorities must inspect, and test water supplies as required by the minister of health to ensure that citizens have access to wholesome water for drinking and domestic purposes. Local authorities must also maintain and secure water sources so that citizens have access to this critical basic service. In ensuring that citizens have access to a safe and adequate water supply, section 64(2) compels premises owners to connect to municipal water. The local authority fixes a minimum charge for this connection to the main water trunks. When the local authority fails to provide adequate water supply to its citizens, provisions are made for subcontracting through the local authorities, who remain with the power to inspect the water supplies. Section 65 places the overall responsibility of water supply in the State as it indicates that water works have to be approved by the State before they commence operations. Such a provision is meant to ensure a supply of safe water which conforms to international standards. Local authorities are then mandated to maintain water supply systems in their areas of jurisdiction.

The lack of water infrastructure and the use of unprotected water sources such as shallow wells are considered a nuisance to public health (Section 85(c)). The section discourages people from drinking such water or using it for domestic purposes as it is considered to be injurious and dangerous to public health. This section thus pays attention to the need to provide safe water that does not compromise human well-being.

3.5.6 Urban Council's Act

The Urban Council's Act (UCA) (Chapter 29:15) is an act of parliament that provides the guidelines and procedures for establishing and administering urban areas in Zimbabwe. The urban areas are categorized as cities, towns, municipalities, councils and town boards. The UCA also confers functions and powers and imposes duties upon urban local authorities to manage affairs in their areas of jurisdiction. Water governance is among the issues addressed in the UCA. Specifically, Part XIII of the UCA provides guidelines and procedures to be followed for urban water governance. The urban local authorities are identified as custodians of the water supply system, a situation showing the decentralization of the water supply system in which local authorities are empowered to provide and maintain a supply of water within or outside the council area. The powers bestowed on urban local authorities signify decentralization efforts in which urban councils are mandated to take care of water supply issues in their areas of jurisdiction. However, most urban councils in the country have been grappling with effectively managing water supply systems due to the rapidly increasing urban population. For example, in Chinhoyi, the daily demand for water supply is 40 ML, yet the municipality of Chinhoyi can only provide 22 ML (Muchetu, 2023). In Harare, the incidences of corruption and embezzlement of public funds have compromised the water supply and revealed the lack of commitment to addressing the water crisis in the city. For example, the US$144.4 million loan given to the City of Harare in 2013 was misused and little was used to rehabilitate and upgrade the water system for Harare, illustrating the lack of commitment on the city's part to improve the water supply (NewsDay Editorial, 2014).

First, the UCA in section 183(1) empowers urban councils to provide and manage water supply within or outside their areas of jurisdiction. Section 183(1)(a) highlights the need for urban local councils to be guided by the Water Act (Chapter 20:22) in providing water supply within and outside their areas of jurisdiction. This provision is critical in showing the alignment and coherence of legislation which is key in water governance as it allows synchronization of different acts and policies. The powers granted to urban councils in section 183 of the UCA have been exercised where urban local authorities have been controlling human settlement development projects ensuring that bulk water infrastructure has been provided before residents can occupy the area. While this has been enforced in most settlements across Zimbabwe, there have been inconsistencies in the application of this provision such that at times some settlements have been established without access to water services. This situation contravenes the envisaged human settlements and also shows the poor water governance practices where local authorities have not been able to fulfil their development mandates. Through section 206 of the UCA, the minister of local government is empowered to direct councils to provide services for residential development. Such services include reticulated water provision. This directive from the minister shows the significance of water and the powers held by the minister to direct urban councils to fulfil their development mandates. However, the political differences have tended to compromise the effective coordination and collaboration between the minister and most urban councils in Zimbabwe. These differences are rooted in the control of the Ministry of Local Government by the ruling ZANU–PF party while urban councils such as Harare and Chitungwiza have been administered by the opposition parties, thereby creating antagonism between the local and central governments, much to the detriment of effective service delivery (Muchadenyika & Williams, 2018). Musemwa (2010) highlighted how the central government blocked and frustrated the initiatives to rehabilitate the water supply in Harare's suburbs, including Mabvuku and Tafara, which had been without water for years. This situation shows the double standards of the central government which prioritizes political support and neglects the right of the citizens to access potable water supply.

Second, section 184 of the UCA highlights that urban councils may compel property owners to connect to a water supply. This section emphasises the significance of water in urban areas and allows the urban councils to 'police' property owners and ensure they are connected to the water system. Implementing these provisions has been difficult in Zimbabwe because the majority of the emerging settlements are not connected to reticulated water infrastructure that should have been installed by the urban councils. Hence, it becomes difficult for the council to police the residents regarding their connections to the main water supply.

Third, councils are tasked to protect water mains in section 186, while section 187 grants urban councils powers to engage in water rationing or restriction schemes. Section 186 addresses a critical issue in water governance on the water infrastructure that needs to be managed and protected. In section 187, the focus on water rationing is a critical aspect of water governance which is often exercised when councils are experiencing water crises. This strategy is considered significant in balancing the water demands against scarce resources and has been implemented in cities such as Cape Town and Johannesburg where water crises have also been escalating over the years (Enqvist & Ziervogel, 2019; Robins, 2019; South African Government News Agency, 2024). However, in Zimbabwe, the water rationing has not been equitably done as some neighbourhoods, especially the low-income neighbourhoods such as Mabvuku, Tafara and Budiriro in Harare, have experienced long spells of water rationing unlike the affluent suburbs of Harare (Human Rights Watch, 2013; Kawadza, 2016). This situation signifies the injustices and marginalization experienced by citizens living in poverty who are often subjected to frequent and longer spells of water rationing.

3.6 Emerging Issues and Conclusion

The legal and regulatory framework governing urban water management in Zimbabwe is anchored by several key laws and regulations that aim to ensure sustainable water use and quality. These laws not only dictate how water resources are managed but also define the roles and responsibilities of different bodies involved in water governance. The cornerstone of Zimbabwe's water legislation is the Water Act, which provides the principal legal framework for the management and allocation of water resources. This act has undergone various amendments to adapt to the changing needs and challenges of water management. It outlines the conditions for obtaining water rights, the use of public water for various purposes, and penalties for the misuse of water. The Water Act emphasizes the State's ownership of all water in Zimbabwe, a principle that centralizes the control of water resources under national government authorities.

Complementing the Water Act, the UCA provides the legal foundation for the governance of cities and towns, including their responsibility for water service delivery. It empowers local councils to create by-laws regarding the provision and management of water services, sewerage systems, and drainage. The act mandates local authorities to ensure adequate water supply and sanitation for their residents, holding them accountable for service quality and efficiency. The Public Health Act is critical in the context of water management as it relates to sanitation and public health. It regulates the conditions under which water and sanitation services must be provided to prevent health hazards. This act reinforces the standards for drinking water quality and sewage treatment, critical aspects of urban water management.

The following are critical issues emerging from the policy analysis for urban water governance in Zimbabwe. First, competing interests between different sectors and stakeholders that is evident in how the central government has continued to be involved in local government issues creating conflicts in water management and planning. The conflicting rationalities and ideologies emerging from these different stakeholders greatly undermine effective urban water governance.

Second, the lack of resources to implement and enforce laws and policies is a great challenge that compromises the success of urban water governance in Zimbabwe. As discussed earlier, local governments are heavily constrained by the financial crisis, largely due to defaulting rates and taxpayers making it difficult to implement and enforce most of the laws and policies on water management.

Third, handling cross-cutting issues is a major challenge, especially in larger cities such as Harare where water issues include livelihood issues such as fishing in waterbodies, urban agriculture and environmental issues such as water pollution and public health issues. Bringing these issues into perspective tends to be a major challenge due to the lack of shared values and common understanding, much to the detriment of good urban water governance.

Fourth, corruption and embezzlement of public funds are a major threat to urban water governance. Corruption erodes public trust and confidence among households, making it difficult for them to participate in water-related matters. Also, the continued involvement of the central government greatly undermines good governance. Lastly, there are great political differences and sabotage by politicians, especially in Bulawayo and Matabeleland where hydropolitics undermines effective water supply planning and management.

References

Adams, E. A., Juran, L., & Ajibade, I. (2018). 'Spaces of exclusion' in community water governance: a feminist political ecology of gender and participation in Malawi's urban water user associations. *Geoforum, 95*, 133–142. https://doi.org/10.1016/j.geoforum.2018.06.016

African Union Commission. (2015). *Agenda 2063: the Africa we want*. Addis Ababa. https://au.int/en/agenda2063/overview

Bhanye, J. I., Mangara, F., Matamanda, A. R., & Kachena, L. (2023). *COVID-19 lockdowns and the urban poor in Harare, Zimbabwe*. Springer. https://doi.org/10.1007/978-3-031-41669-9

Brooke, A., & Fenner, R. A. (2023). Improving urban water management and building water supply resilience in the city of Harare, Zimbabwe—A systems view. *Civil Engineering and Environmental Systems, 40*(4), 195–228. https://doi.org/10.1080/10286608.2023.2233910

Chamboko, T., Mukhebi, A. W., Callaghan, C. J. O., Peter, T. F., Kruska, R. L., Medley, G. F., Mahan, S. M., & Perry, B. D. (1999). The control of heartwater on large-scale commercial and smallholder farms in Zimbabwe. *Preventive Veterinary Medicine, 39*(3), 191–210. https://doi.org/10.1016/S0167-5877(98)00144-5

Chikozho, C., & Latham, J. (2005). Shona customary practices in the context of water sector reforms in Zimbabwe. International workshop on African Water Laws: Plural Legislative Frameworks for Rural Water Management in Africa, 26–28 January 2005, Johannesburg, South Africa. https://assets.publishing.service.gov.uk/media/57a08ca8ed915d622l c0014a7/R8323-CHIKOZHO-C.pdf

City of Harare. (2020). *Harare voluntary local review of Sustainable Development Goals (SDGs) report, June 2020*. City of Harare.

Constitution Select Committee. (2012). Milestones towards a new constitution for Zimbabwe. Overview of the constitution making process. https://constitutionnet.org/sites/default/files/zimbabwe-overview_of_the_constitution_making_process.pdf

DeGhetto, K., Gray, J. R., & Kiggundu, M. N. (2016). The African Union's agenda: aspirations, challenges, and opportunities for management research. *Africa Journal of Management, 2*(1), 93–116. https://doi.org/10.1080/23322373.2015.1127090

Di Vaio, A., Trujillo, L., D'Amore, G., & Palladino, R. (2021). Water governance models for meeting sustainable development goals: A structured literature review. *Utilities Policy, 72*, Article 101255. https://doi.org/10.1016/j.jup.2021.101255

Enqvist, J. P., & Ziervogel, G. (2019). Water governance and justice in Cape Town: An overview. *Wiley Interdisciplinary Reviews Water, 6*(4), Article e1354. https://doi.org/10.1002/wat2.1354

Gogo, J. (2019). Should residents continue paying for water? *The Herald*. 6 May. https://www.herald.co.zw/should-residents-continue-paying-for-water/. Accessed 20 December 2024.

Government of Zimbabwe. (2012). *Zimbabwe national water policy*. Government of Zimbabwe.

Government of Zimbabwe. (2020). *Zimbabwe progress review report of sustainable development goals (SDGs)*, December 2020. Revised edition. Government of Zimbabwe.

Government of Zimbabwe. (2021). *Zimbabwe's second voluntary national review (VNR)*. Government of Zimbabwe.

Hadebe, S., Mandaza, I., Moyo, G., Mutondoro, F., & Ncube, M. J. (2015). *Annual state of corruption report: Focus on state-owned enterprises*. Transparency International Zimbabwe.

Human Rights Watch. (2013). *Troubled Water: Burst pipes, contaminated wells, and open defecation in Zimbabwe's capital*. https://www.hrw.org/report/2013/11/19/troubled-water/burst-pipes-contaminated-wells-and-open-defecation-zimbabwes. Accessed 8 December 2024.

Kadungure. (2013). Chombo directive on writing off council debt 'irrational' says MDC-T. *Nehanda Radio*. 25 July. https://nehandaradio.com/2013/07/25/chombo-directive-on-writing-off-council-debt-irrational-says-mdc-t/. Accessed 7 November 2024.

Kawadza, S. (2016). Water woes dog Harare residents. *The Herald*. 15 April. https://www.herald.co.zw/water-woes-dog-harare-residents/. Accessed 17 December 2024.

Magadza, C. H. D. (2003). Lake Chivero: a management case study. *Lakes & Reservoirs: Research & Management, 8*(2), 69–81. https://doi.org/10.1046/j.1320-5331.2003.00214.x

Manjengwa, F., Nhiwatiwa, T., Nyakudya, E., & Banda, P. (2019). Fish from a polluted lake (Lake Chivero, Zimbabwe): A food safety issue of concern. *Food Quality and Safety, 3*(3), 157–167. https://doi.org/10.1093/fqsafe/fyz015

Manzungu, E., & Dzingirai, V. (2012). Towards empowered stakeholder participation in water resource management in Zimbabwe. *Journal of Social Development in Africa, 27*(1), Article 85.

Mapedza, E., Manzungu, E., Rosen, T., Ncube, P., & Van Koppen, B. (2016). Decentralised water governance in Zimbabwe: Disorder within order. *Water Resources and Rural Development, 8*, 1–11. https://doi.org/10.1016/j.wrr.2016.05.001

Maponga, G. (2013). Municipalities write off water bills. *The Herald*. 31 July. https://www.herald.co.zw/municipalities-write-off-water-bills/. Accessed 7 December 2024.

Maramura, T. C. (2018). The efficacy of prepaid water meters for potable water service provision in the Harare City Council, Zimbabwe. (Unpublished doctoral dissertation, North-West University, Potchefstroom, South Africa).

Marcuse, P. (2010). Spatial justice: derivative but causal or social justice. In B. Bret, P. Gervais-Lambony, & C. P. Hancock (Eds.), *Justices Et Injustices Spatiales* (pp. 76–92). Presses Universitaires de Paris Quest.

Maruta, E., Matsa, M., Mupepi, O., & Dzawanda, B. (2023). Water safety and associated public health impacts in upcoming and informal urban settlements in Zimbabwe: Case of Amsterdam Park in Harare. *Sustainable Water Resources Management, 9*(5), Article 138. https://doi.org/10.1007/s40899-023-00921-3

Mashingaidze, T. M. (2019). The Kariba Dam, discursive displacements and the politics of appropriating a waterscape in Zimbabwe, 1950s-2017. *LIMINA: A Journal of Historical and Cultural Studies, 25*(1), 1–15. https://search.informit.org/doi/10.3316/informit.087122106039661

Masimba Nyama, G., & Walnycki, A. (2024). *Harare: City report (August 26, 2024)*. https://doi.org/10.2139/ssrn.4943897

Matamanda, A. R. (2022). Are we there yet? Prospects and barriers to implementing smart city initiatives in Harare, Zimbabwe. In C. Chavunduka, W. Timo de Vries, & P. D. Diaz (Eds.), *Sustainable and smart spatial planning in Africa: Case studies and solutions* (pp. 107–119). CRC Press.

Matamanda, A. R., Mphambukeli, T. N., & Chirisa, I. (2024). Exploring water-gender-health nexus in human settlements in Hopley, Harare. *Cities & Health, 8*(1), 44–58. https://doi.org/10.1080/23748834.2022.2136557

Matsa, M., Mavugara, R., & Dzawanda, B. (2021). Urban domestic water supply crisis in the city of Gweru, Zimbabwe. *GeoJournal, 86*, 1173–1192. https://doi.org/10.1007/s10708-019-10118-x

Muchadenyika, D., & Williams, J. J. (2018). Politics, centralization and service delivery in urban Zimbabwe. *Journal of Southern African Studies, 44*(5), 833–853. https://doi.org/10.1080/03057070.2018.1500748

Muchetu, E. (2023). Debilitating water challenges throttling Chinhoyi. *Suburban*. 07 July. https://www.suburban.co.zw/debilitating-water-challenges-throttling-chinhoyi/. Accessed 19 December 2024.

Mukonavanhu, T., Ukwandu, D., & Nel-Sanders, D. (2021). Commercialisation of water supply in Zimbabwe and its effect on the poor: a working framework. *Africa's Public Service Delivery and Performance Review, 9*(1), Article a536. https://doi.org/10.4102/apsdpr.v9i1.536

Musemwa, M. (2006). Disciplining a 'dissident' city: Hydropolitics in the city of Bulawayo, Matabeleland, Zimbabwe, 1980-1994. *Journal of Southern African Studies, 32*(2), 239–254. https://doi.org/10.1080/03057070600656119

Musemwa, M. (2008). Early struggles over water: from private to public water utility in the city of Bulawayo, Zimbabwe, 1894–1924. *Journal of Southern African Studies, 34*(4), 881–898. https://www.jstor.org/stable/40283198

Musemwa, M. (2010). From 'Sunshine City' to a landscape of disaster. The politics of water, sanitation and disease in Harare, Zimbabwe, 1980-2009. *Journal of Developing Societies, 26*(2), 165–206. https://doi.org/10.1177/0169796X1002600202

Musemwa, M. (2021). Urban struggles over water scarcity in Harare. *Daedalus, the Journal of the American Academy of Arts & Sciences, 150*(4), 27–47. https://doi.org/10.1162/daed_a_01871

Muyambo, T., & Maposa, R. S. (2014). Linking culture and water technology in Zimbabwe: Reflections on Ndau experiences and implications for climate change. *Journal of African Studies and Development, 6*(2), 22–28. https://doi.org/10.5897/JASD2013.0266

Ndoziya, A. T., Hoko, Z., & Gumindoga, W. (2019). Assessment of the impact of pit latrines on groundwater contamination in Hopley Settlement, Harare, Zimbabwe. *Journal of Water, Sanitation and Hygiene for Development, 9*(3), 464–476. https://doi.org/10.2166/washdev.2019.179

NewsDay Editorial. (2014). Harare City Council loan misuse an act of corruption. *The NewsDay*. 23 May. https://www.newsday.co.zw/opinion-amp-analysis/article/133457/harare-city-council-loan-misuse-an-act-of-corruption. Accessed 19 December 2024.

Nyabeze, T. H. (2015). *Progressive reform in the new constitution of Zimbabwe: A balance between the preservation and transformative constitution making process*. Konrad-Adenauer-Stifung e.V.

Pieterse, E. (2019). Urban governance and spatial transformation ambitions in Johannesburg. *Journal of Urban Affairs, 41*(1), 20–38. https://doi.org/10.1080/07352166.2017.1305807

Reniko, G., & Kolawole, O. D. (2020). 'They don't read metres, they only bring bills': Issues surrounding the installation of prepaid water metres in Karoi town, Zimbabwe. *South African Geographical Journal, 102*(3), 356–371. https://doi.org/10.1080/03736245.2019.1691046

Robins, S. (2019). 'Day zero', hydraulic citizenship and the defence of the commons in Cape Town: A case study of the politics of water and its infrastructures (2017-2018). *Journal of Southern African Studies, 45*(1), 5–29. https://doi.org/10.1080/03057070.2019.1552424

Royo, M. G., Diep, L., Mulligan, J., Mukanga, P., & Parikh, P. (2022). Linking the UN sustainable development goals and African agenda 2063: Understanding overlaps and gaps between the global goals and continental priorities for Africa. *World Development Sustainability, 1*(2022), Article 100010. https://doi.org/10.1016/j.wds.2022.100010

Schulz, C. (2019). Governance-related values as dimensions of good water governance. *Wiley Interdiscip Rev Water, 6*(1), e1322. https://doi.org/10.1002/wat2.1322

Shonhe T. (2011). Public participation in the constitution making process in Zimbabwe (2009–2010). (Unpublished master's thesis, University of Pretoria, Pretoria, South Africa).

Soja, E. W. (2009). The city and spatial justice. *Justice Spatiale/Spatial Justice, 1*(1), 1–5. https://www.jssj.org/article/la-ville-et-la-justice-spatiale/?lang=en

Soja, E. W. (2010). The city and spatial justice. In B. Bret, P. Gervais-Lambony, C. Hancock, & F. Landy (Eds.), *Justice Et Injustices spatiales* (pp. 55–75). Presses Univeritaires de Paris Quest.

South African Government News Agency. (2024). *City of Johannesburg to implement water restrictions.*. https://www.sanews.gov.za/south-africa/city-johannesburg-implement-water-restrictions. Accessed 17 December 2024.

Taruvinga, G. R. (2024). Zimbabwe's water supply is in crisis: how political interference and underfunding did the damage. *The Conversation*. 23 October. https://theconversation.com/zimbabwes-water-supply-is-in-crisis-how-political-interference-and-underfunding-did-the-damage-239864. Accessed 21 December 2024.

Tom, T., & Munemo, E. (2015). Republic of Zimbabwe National Water Policy: A desk review of the gaps between policy and its implementation. *International Journal of Public Policy and Administration Research, 2*(3), 60–72. https://doi.org/10.18488/journal.74/2015.2.3/74.3.60.72

United Nations. (2017). New Urban Agenda. *Habitat III*. https://habitat3.org/wp-content/uploads/NUA-English.pdf

United Nations. Department of Economic and Social Affairs. (2015). The 17 goals. https://sdgs.un.org/goals

Worster, D., & Empire, R. O. (1985). Water, aridity and the growth of the American west. In *Introduction to Rivers of Empire: water, aridity and the growth of the American west* (pp. 19–61). Oxford University Press.

ZimEye. (2021). Zinwa in US$100 million Chinese bogus water tender scandal. ZimEye., 7 August. https://www.zimeye.net/2021/08/07/zinwa-in-us100-million-chinese-bogus-water-tender-scandal/. Accessed 17 December 2024.

Zinyuke, R. (2024). Lake Chivero ecological disaster … health scare as wildlife dies. *The Herald*. 17 December. https://www.herald.co.zw/lake-chivero-ecological-disaster-health-scare-as-wildlife-dies/. Accessed 17 December 2024.

Chapter 4
Institutionalism and Urban Water Governance in Zimbabwe

Abstract This chapter critically examines the urban water governance landscape in Zimbabwe through the lens of institutionalism. It explores the complex interplay between institutional structures, rules, norms, and routines that shape the management and distribution of water resources in urban settings. The analysis highlights significant challenges such as water scarcity, infrastructural inadequacies, governance failures exacerbated by political and economic instability, and the emergence of water mafias. Despite Zimbabwe's abundant natural water sources, urban areas such as Harare and Bulawayo grapple with recurring water crises due to inadequate investment in infrastructure and inefficient governance frameworks. To address these issues, the chapter proposes the establishment of adaptive, transparent, and inclusive institutions capable of engaging in cross-scalar governance and partnerships. These institutions should be designed to adapt to changing conditions such as population growth, climate variability, and technological advancements, ensuring continuous learning and policy adaptation. They must also facilitate broad stakeholder engagement, including marginalized and vulnerable communities, to ensure that governance frameworks are equitable and comprehensive. The chapter discusses the roles of key institutions, including governmental bodies, local communities, NGOs, the private sector, and academic institutions, and addresses the detrimental impact of water mafias, which manipulate water access for economic and political gain. These institutions can improve sustainability and access to clean water in Zimbabwe's urban centers by promoting hydrocitizenship and innovations such as smart water management systems. The chapter offers insights into how strengthened institutional frameworks could enhance the management and sustainability of water resources in Zimbabwe's urban areas, providing a roadmap for reforming water governance systems to improve sustainability and access to clean water.

Keywords Urban water governance · Institutionalism · Zimbabwe · Hydrocitizenship · Infrastructure · Decentralization · Stakeholder engagement

A. R. Matamanda et al., *Urban Water Governance in Postcolonial Zimbabwe*, Local and Urban Governance, https://doi.org/10.1007/978-3-031-96973-7_4

4.1 Introduction

Water is essential for sustaining life and supporting socioeconomic development (Goswami & Bisht, 2017). Urban regions globally encounter significant water governance challenges due to factors such as population growth, industrialization, and environmental constraints (Bakker, 2013). These challenges are notably acute in Southern African countries such as Zimbabwe, characterized by water scarcity intensifying urban developmental complexities. Zimbabwe's urban centers, including Harare, the capital city, and Bulawayo, the second largest city in the country, face severe difficulties in securing sustainable water supplies amid expanding populations and inadequate infrastructure investments. In Zimbabwe, urban water scarcity is compounded by complex governance issues (Mapedza et al., 2016). Despite the availability of natural water sources, the supply is often erratic and insufficient to meet the demands of the urban populace. Recurring water crises in these cities reveal the failures of existing management systems to effectively address climatic variability challenges, rapid urbanization, and infrastructural degradation (Taruvinga, 2024a, b). This situation highlights the need for a thorough examination of the governance frameworks that regulate water resources.

Water governance encompasses the political, social, economic, and administrative systems in place to develop and manage water resources and the delivery of water services at different levels of society (Woodhouse & Muller, 2017). In urban settings, effective water governance is crucial due to the dense population and the substantial demand for sustainable water delivery and sanitation services (Matamanda et al., 2024). The complexity of urban water management requires coordinated actions among multiple stakeholders, adherence to regulations, and the implementation of policies that ensure equitable access to clean water (Mgwele et al., 2024). This is especially important in cities where infrastructure may be underdeveloped or ageing, and where the risk of water scarcity and contamination can significantly affect public health and socioeconomic stability (Rammile et al., 2024).

Zimbabwe is characterized by a combination of high plateau and mountainous regions in the east, influencing the natural distribution and availability of water resources. The country's climate varies from tropical to subtropical, making it prone to periods of drought and erratic rains in most regions (Mamombe et al., 2017). Economically, Zimbabwe has faced numerous challenges over the past decades, including hyperinflation, unemployment, and a decline in industrial output, which have strained public services and infrastructure, including those for water governance (Bhanye et al., 2023). The socioeconomic landscape, marked by a significant portion of the population living in poverty, exacerbates challenges in urban water management as resources are often not sufficient to meet the needs of the urban populace. The migration from rural to urban areas has put additional pressure on existing water management systems, highlighting the urgent need for effective governance.

This chapter explores the dynamics of institutionalism, understood as the study of the structures, rules, norms, and routines that govern social, political, and economic interactions, within the context of urban water governance in Zimbabwe. Institutionalism emphasizes how formal organizations, informal practices, and established institutional arrangements influence decision-making and resource management (Peters, 2022). It dissects the roles and impacts of various institutions and governmental bodies to local community structures—in shaping water governance. The chapter provides a comprehensive analysis of how institutional practices and reforms influence the efficacy of water governance in urban settings. Additionally, it covers aspects of institutionalism, such as the impact of political stability, policy implementation, and the interplay between formal and informal institutions on water management. The chapter offers insights into how strengthened institutional frameworks could enhance the management and sustainability of water resources in Zimbabwe's urban areas.

4.2 Theoretical Framework

4.2.1 Institutionalism

Institutionalism in governance is a broad theoretical approach that studies how institutions—both formal and informal—shape the behavior of individuals and the functioning of societies (Peters, 2022). Institutions, defined as established laws, norms, and practices that guide social behavior, play a crucial role in determining how public policies are developed, implemented, and sustained over time (North, 1990). In the context of public policy and governance, institutionalism provides a framework for understanding the complexity and durability of political structures, the role of bureaucracy, and the interactions between different governance entities (Koliba et al., 2017). This approach is particularly pertinent to the governance of common resources like water, where multiple stakeholders and competing interests must be balanced within established regulatory frameworks (Ostrom, 1990).

There are different perspectives on institutionalism. Historical institutionalism emphasizes the path-dependent nature of institutions, suggesting that choices made during critical junctures in history exert a lasting influence on present and future policy outcomes (Pierson, 2004; Thelen, 1999). This perspective is valuable in understanding long-standing water governance structures in Zimbabwe, where colonial legacies have shaped current institutional frameworks and practices. Sociological institutionalism focuses on the role of culture, norms, and cognitive scripts in the creation and maintenance of institutions (March & Olsen, 1989). It examines how societal values and norms influence policy decisions and institutional efficacy, particularly relevant in water governance, which involves cultural perceptions of water as a communal versus a marketable resource. New institutionalism, or rational choice institutionalism, posits that individuals are rational actors who make decisions based on their assessments of costs and benefits within the constraints imposed by institutions (Shepsle, 1989). This model applies to water governance, where policymakers and stakeholders make strategic decisions based on economic and social incentives within regulatory frameworks.

Thus, several key principles underpin institutional theory, providing a foundation to analyze governance structures. Table 4.1 shows the key principles of institutional theory and their application in Zimbabwe's urban water governance.

These principles are critical for analyzing the layers of governance and understanding the nuances of institutional interactions in the management of urban water resources. Through these theoretical insights, the subsequent analysis of Zimbabwe's urban water governance structures will identify the institutional barriers and enablers that affect water service delivery and sustainability.

4.2.2 *Application of Institutional Theory to Water Governance*

The application of institutional theories to water governance can be seen in the ways policy frameworks and institutional arrangements are designed, maintained, and reformed. For instance, historical institutionalism helps explain why certain outdated and inefficient water governance practices persist in Zimbabwe, as institutions tend to resist changes due to path dependency and the vested interests of established elites (Bates, 2005). Sociological institutionalism offers insights into

Table 4.1 Key principles of institutional theory and their application in Zimbabwe's urban water governance

Key principles	Description	Relevance to Zimbabwe
Path dependency	Historical paths influence decisions and strategies within institutions. Established courses of action become hard to change, creating persistent patterns (Thelen, 1999)	The legacy of colonial water management systems continues to shape current strategies and challenges, emphasizing the need for historical context in reforms
Institutional compatibility	The degree to which various institutions support and reinforce each other enhances governance effectiveness through coherent and aligned policies across different levels and sectors (Li et al., 2018)	Misalignment between local water management practices and national policies can lead to conflicts and inefficiencies in service delivery
Institutional isomorphism	Institutions converge over time through coercive, mimetic, and normative mechanisms, as defined by DiMaggio and Powell (1983)	Urban water governance may see municipalities adopting similar management technologies or policies due to regulatory pressures or peer successes
Adaptive efficiency	The flexibility of institutions to evolve and adapt in response to changing circumstances—a concept introduced by North (1990)	Crucial for managing complex and dynamic urban water systems in Zimbabwe, especially in adapting to unpredictable events like droughts or technological advancements

how communal norms around water usage influence the success or failure of water policies, especially in rural areas where community management of water resources is common (Cleaver, 1999). Lastly, rational choice institutionalism can be applied to understand the decisions behind water pricing and allocation policies, reflecting the rational calculations by different stakeholders aiming to maximize their utilities within the constraints of governmental policies and regulations (Tang, 1992).

The functionality of water governance systems often hinges on the robustness and coherence of the institutions involved. Effective institutions are characterized by clear mandates, strong regulatory frameworks, and mechanisms that ensure accountability and transparency. In the context of water governance, functionality can be assessed through the efficiency of water distribution, equitable access to water resources, and the sustainability of water use practices. Institutionalism helps to highlight how regulatory bodies, policy frameworks, and management practices interact to produce these outcomes. For example, studies by Ostrom (1990) have demonstrated that well-defined property rights, local enforcement of resource management norms, and community participation in decision-making processes are crucial for the sustainable management of common-pool resources, including water. These elements foster cooperation among users and help to prevent the overexploitation of resources.

Conversely, dysfunctions in water governance systems often arise from institutional failures, which may include weak enforcement of laws, lack of coordination among agencies, and inadequate stakeholder engagement. Institutional mismatches, where there is a misalignment between the scale of environmental problems and the institutions addressing them, can also lead to governance failures. For instance, when local governments lack the authority or resources to effectively implement national water policies, it can result in fragmented and inefficient water management. Moreover, path dependency, a key concept in institutionalism, highlights how historical decisions continue to shape current governance practices, sometimes leading to persistent inefficiencies. In many urban areas in developing countries, colonial-era water infrastructure and management practices have not been adequately updated to meet current demands, leading to systemic failures in water delivery and sanitation services (Saleth & Dinar, 2004).

4.3 Institutional Structures and Urban Water Governance in Zimbabwe

4.3.1 Governmental Bodies and Agencies

The institutional framework for water management in Zimbabwe is complex, involving multiple governmental bodies and agencies, each with distinct roles and responsibilities. Understanding these entities and their interactions is crucial for understanding the effectiveness of urban water management.

4.3.2 Ministry of Environment, Water and Climate

The Ministry of Environment, Water, and Climate plays a pivotal role in shaping and overseeing water governance in Zimbabwe. As the primary governmental body responsible for water management at the national level, the ministry is tasked with formulating policies, establishing regulatory standards, and coordinating activities across various sectors related to water resources, environmental protection, and climate resilience. Its central mandate is to ensure that water governance aligns with national development priorities, environmental sustainability, and international best practices. A key function of the ministry is legislative development, which involves drafting and enacting laws that regulate the use, allocation, and conservation of water resources. These laws provide the framework within which water management institutions and stakeholders operate, ensuring that water usage is equitable and sustainable. Additionally, the ministry is responsible for policy formulation, setting the strategic direction for the country's water governance. This includes addressing emerging challenges such as water scarcity, pollution, and the impacts of climate change. By developing comprehensive policies, the ministry seeks to balance the needs of various sectors, including agriculture, industry, and urban development, while safeguarding the environment (Government of Zimbabwe, 2017).

The Ministry of Environment, Water, and Climate also plays a crucial role in international cooperation on water-related issues. Zimbabwe is a member of several regional and international agreements and organizations focused on water governance, such as the Southern African Development Community Protocol on Shared Watercourses. The ministry represents the country in these platforms, ensuring that Zimbabwe's interests are protected and that it adheres to international commitments. Through these engagements, the ministry facilitates knowledge exchange, capacity building, and access to funding for water-related projects.

In carrying out its mandate, the ministry collaborates closely with other governmental bodies to integrate water management with broader environmental and economic policies. For example, it works with the Ministry of Agriculture to address the water needs of the agricultural sector, which is a critical driver of Zimbabwe's economy. Similarly, it engages urban planning and local government authorities to ensure that water management is incorporated into city planning and development strategies. This cross-sectoral approach underscores the ministry's recognition of water as a vital resource that underpins multiple aspects of national development.

Despite its significant role, the Ministry of Environment, Water, and Climate faces several challenges in fulfilling its mandate. Limited financial and technical resources often constrain the ministry's ability to implement policies and to effectively enforce regulations. Additionally, the ministry operates in the context of competing priorities and limited coordination among various stakeholders, which can lead to inefficiencies in water governance. Addressing these challenges requires strengthening institutional capacity, enhancing inter-ministerial collaboration, and mobilizing resources to support the ministry's activities (Government of Zimbabwe, 2017).

4.3.3 Zimbabwe National Water Authority

ZINWA is an important institution in Zimbabwe's water governance framework, tasked with the critical responsibility of managing the nation's water resources. Established through the Water Act of 1998, ZINWA was created to centralize water resource management, addressing inefficiencies and improving coordination across the country's water systems. Operating under the Ministry of Environment, Water, and Climate, ZINWA plays an instrumental role in planning, developing, and sustaining water resources to meet national needs while ensuring equitable access (Mutandwa & Vyas-Doorgapersad, 2023a, b).

One of ZINWA's primary responsibilities is the planning and development of water resources to ensure their sustainable use. This involves conducting assessments of water availability and demand across the country and identifying strategies to optimize the use of surface and groundwater resources. By overseeing the construction and maintenance of major dams and reservoirs, ZINWA ensures that water is stored effectively to support agriculture, industry, and domestic consumption, particularly in drought-prone areas (Mazivarimwe, 2020). Groundwater management is another key focus, with the authority tasked with monitoring and regulating the extraction of this vital resource to prevent overuse and contamination.

ZINWA also plays a vital role in maintaining and upgrading water infrastructure, including pipelines, treatment plants, and irrigation systems. These systems are essential for ensuring that both urban and rural areas have access to safe and reliable water supplies (Dhoba, 2022). The authority is responsible for the operation of national water treatment facilities, which are crucial for delivering potable water that meets national and international quality standards. This includes routine testing and monitoring to detect and address pollutants, ensuring the safety of water for consumption and other uses.

Equitable water allocation is a core component of ZINWA's mandate, ensuring that water resources are distributed fairly across different sectors and regions. The authority works to balance the competing demands of agriculture, industry, and domestic users while prioritizing access for vulnerable and marginalized communities. This commitment to equity is reflected in ZINWA's collaboration with local authorities and community stakeholders to align national water management strategies with local needs and conditions. Working directly with municipalities and rural councils, ZINWA helps to implement policies that reflect the realities of local water challenges and opportunities.

Despite its extensive responsibilities, ZINWA faces significant challenges in fulfilling its mandate. The authority often struggles with inadequate financial and technical resources, limiting its ability to maintain infrastructure and expand services (Mapedza et al., 2016). Aging infrastructure, particularly in rural areas, exacerbates these difficulties, leading to inefficiencies and water losses. Additionally, ZINWA operates within a complex institutional environment, where overlaps and conflicts with other water management entities, such as catchment councils and municipal authorities, can impede coordination and effectiveness.

To address these challenges, ZINWA has increasingly embraced partnerships with international organizations, private sector entities, and NGOs. These collaborations have facilitated access to funding, technical expertise, and innovative technologies that enhance the authority's capacity to deliver on its mandate. For example, ZINWA has been involved in projects aimed at modernizing irrigation systems and introducing real-time water quality monitoring tools.

4.3.4 Local Authorities

Local authorities, which include city and town councils, are also pivotal in urban water governance. They are responsible for water distribution and sanitation services within their jurisdictions. These bodies work in collaboration with ZINWA, especially in large urban centers like Harare and Bulawayo, where the water demand is high. Local authorities manage the local water supply systems, sewage, and waste treatment facilities. They are tasked with ensuring that residents have access to safe and reliable water, often facing challenges related to ageing infrastructure and limited financial resources (Musemwa, 2021).

Despite their critical role, local authorities in Zimbabwe face numerous challenges that impede their ability to provide reliable water services (Musemwa, 2021). One of the major issues is the ageing infrastructure, much of which has deteriorated due to a lack of maintenance and investment. This problem is exacerbated by financial constraints, as many water departments operate under tight budgets that do not allow for significant upgrades or expansions (Brooke & Fenner, 2023).

Another significant challenge is the mismatch between water supply and demand. Urban population growth has outpaced the capacity of existing water infrastructure, leading to frequent shortages and rationing in many cities (Musemwa, 2021). Additionally, bureaucratic inefficiencies and corruption have further hindered effective water management. Instances of mismanagement and lack of accountability have led to public distrust and non-payment of water bills, which in turn reduces the revenue available for reinvestment in the water sector (Hove & Tirimboi, 2011). Local authorities also struggle with technical challenges, such as water loss through leaks and illegal connections. By inflating operational costs and reducing the amount of billable water, these technical problems not only waste limited water resources but also make water departments less financially viable.

Effective governance is further complicated by coordination issues between ZINWA and local authorities. While ZINWA controls the bulk water supply, local authorities manage the distribution within cities, which sometimes leads to conflicts over jurisdiction and resource allocation (Muchadenyika & Williams, 2018). Effective collaboration and clear delineation of responsibilities are essential to improve governance and ensure the sustainability of urban water systems in Zimbabwe.

4.3.5 Catchment Councils

Zimbabwe is divided into several water catchment areas, each managed by a catchment council. These councils are part of a decentralized approach to water management intended to involve local stakeholders directly in the decision-making process (Mapedza et al., 2016). Catchment councils are responsible for managing water resources at a catchment level, including issuing water permits, overseeing water usage, and implementing catchment management plans (Derman & Manzungu, 2016). One of the core functions of catchment councils is issuing water permits, which allocate usage rights to various sectors, such as agriculture, industry, and domestic users. This regulated approach helps prevent overexploitation and promotes equitable access to water resources. In addition to issuing permits, the councils monitor water usage to ensure compliance with regulations and prevent unauthorized extraction or pollution. A strategic focus is also placed on developing and implementing catchment management plans, which serve as blueprints for sustainable water management by considering environmental needs, stakeholder input, and future water demands (Derman & Manzungu, 2016).

Catchment councils play a vital role in balancing the competing needs of diverse water users. Engaging local communities, water users, and other stakeholders, they create a platform for inclusive governance that addresses the unique challenges of each catchment area. This localized approach helps promote equity, particularly for marginalized and vulnerable groups who often face barriers to accessing water. Furthermore, by fostering the adoption of sustainable water use practices, councils contribute to long-term resource conservation and environmental protection. Their proximity to local stakeholders also enhances transparency and accountability, making water management decisions more responsive and grounded in local realities (Mapedza et al., 2016).

Despite their importance, catchment councils face significant challenges that hinder their effectiveness. Resource constraints, including inadequate financial and technical support, often limit their capacity to perform critical functions such as monitoring and enforcement (Mapedza et al., 2016). Additionally, many councils struggle with insufficient training and expertise among members, which undermines the development and execution of effective management plans. Stakeholder conflict also poses a persistent challenge, as balancing the competing demands of different water users can be contentious and complex (Derman & Manzungu, 2016). These issues are further exacerbated by the impact of climate change, such as erratic rainfall patterns and increased water scarcity, which add layers of unpredictability to water resource management. Institutional overlaps and ambiguities in roles and responsibilities also undermine the effectiveness of catchment councils. For instance, unclear delineations between the functions of the councils and ZINWA can lead to inefficiencies and governance conflicts. Such overlaps highlight the need for greater policy harmonization and institutional clarity to streamline water governance processes. Table 4.2 summarizes the key roles and responsibilities within water management, including the specific functions of ZINWA, local authorities and the Ministry of Environment, Water and Climate.

Table 4.2 Roles and responsibilities of key stakeholders in water management

Entity	Role	Responsibilities
ZINWA	National water management	Strategic planning and development of water resources Maintenance of water infrastructure Water allocation Water quality management
Local authorities	Local water service management	Implementation of national water policies at the local level Maintenance of local water supply systems Coordination with ZINWA on water supply issues
Ministry of Environment, Water and Climate	Policy oversight and regulation	Oversight of ZINWA activities Development of national policies on water, environment, and climate Regulation of standards and practices in water management
Catchment councils	Regional water management	Management of water resources within specific catchments Issuing of water permits and monitoring usage Implementing catchment management plans Engaging local stakeholders in water management decisions

In Zimbabwe, the management of urban water resources is not solely the domain of government agencies. NGOs, CSOs, the private sector and resident associations also play crucial roles, contributing to various aspects of water governance, ranging from policy advocacy to direct service delivery and community engagement.

4.3.6 Non-governmental organizations

NGOs in the water sector often focus on areas where government services are lacking, especially in underserved urban and peri-urban areas. These organizations frequently engage in building and rehabilitating water infrastructure, such as boreholes and sanitation facilities, to improve access to clean water (Van Welie & Romijn, 2018). Additionally, they play a significant role in disaster response by providing emergency water supplies during droughts or outbreaks of waterborne diseases. NGOs like WaterAid Zimbabwe and the International Water Management Institute have been instrumental in projects that aim to enhance water quality and availability in urban areas. Table 4.3 shows the roles and responsibilities of NGOs in water management in Zimbabwe.

4.3.7 Civic Society Organizations

CSOs include a broad range of organizations, such as community-based organizations, residents' associations, and environmental groups. They are deeply involved in advocating for the rights of communities to access clean and affordable water (Bandauko, 2015). CSOs also monitor government actions and push for

Table 4.3 Roles and responsibilities of NGOs in water management

Role	Responsibilities	Example
Infrastructure development	Constructing and maintaining water supply systems	Ensuring a reliable and sustainable water supply infrastructure
Capacity building	Training local communities and water management committees in sustainable water use and maintenance practices	Empowering local stakeholders to manage and conserve water effectively
Advocacy and policy influence	Working to influence water governance policies to ensure they are equitable and sustainable NGOs often advocate for integrating climate change adaptation	Advocating for climate-resilient water policies
Research and development	Conducting research to better understand and address specific water management challenges, such as groundwater depletion or pollution	Researching groundwater issues and pollution control

Table 4.4 Roles and responsibilities of CSOs in community-based water management in Zimbabwe

Role	Responsibilities	Example
Community mobilization and empowerment	Engaging communities to take an active role in managing their water resources, including education on water use rights and conservation	Facilitates community-led water management initiatives
Monitoring and evaluation	Keeping track of government and NGO projects to ensure they meet community needs and are effectively implemented	Ensures transparency and accountability in water projects
Conflict resolution	Acting as mediators in disputes over water resources, which can occur between different user groups within urban areas	Resolves conflict to ensure fair and sustainable water distribution
Environmental advocacy	Promoting sustainable water management practices and conservation efforts, focusing on protecting urban watersheds and ecosystems	Advocacy against industrial pollution and urban sprawl

transparency and accountability in the management of water resources. Table 4.4 shows the roles and responsibilities of CSOs in community-based water management in Zimbabwe.

The contributions of NGOs and CSOs are vital in filling gaps left by governmental inadequacies, especially in resource-limited settings. They bring innovation, flexibility, and often a closer connection to community needs. However, their efforts can be limited by several factors, including resource constraints, scalability and sustainability, and political and institutional barriers. Many NGOs and CSOs operate with limited budgets and are heavily reliant on external funding, which can be unstable and unpredictable. Projects initiated by NGOs and CSOs may face challenges in scaling up and ensuring long-term sustainability without continuous external support. In some cases, these organizations may face opposition from local authorities or national governments, particularly if their activities challenge existing power structures or expose corruption. Despite these challenges, the role of NGOs and CSOs remains indispensable in promoting effective, equitable, and sustainable urban water management in Zimbabwe. Their ongoing efforts to complement governmental functions underscore the need for a collaborative approach that leverages the strengths of all sectors to address the complex challenges of urban water governance.

4.3.8 Private Sector Involvement

The private sector plays a significant role in urban water governance in Zimbabwe, complementing governmental and institutional efforts to effectively manage water resources. Public institutions often face financial and technical limitations, making private sector involvement a valuable strategy for improving water service delivery, infrastructure development, and resource management (Musemwa, 2021). Public–private partnerships and direct private sector participation have emerged as key mechanisms to address the growing challenges of urban water management (Lima et al., 2021). One of the primary contributions of the private sector is in the development and maintenance of water infrastructure. Private entities have provided critical financing and expertise to construct and upgrade water treatment plants, pipelines, and reservoirs. Some of the urban areas in Zimbabwe have benefited from public–private partnership models, where private firms collaborate with public institutions to expand and modernize water supply systems (Moyo, 2013; Mutandwa & Vyas-Doorgapersad, 2023a, 2023b). These partnerships not only bring in much-needed investment but also introduce advanced technologies to enhance the efficiency and reliability of urban water infrastructure. Table 4.5 shows the roles and responsibilities of the private sector in urban water governance in Zimbabwe.

Table 4.5 Roles and responsibilities of the private sector in urban water governance in Zimbabwe

Role	Responsibilities
Infrastructure development	Financing and constructing new water treatment plants, pipelines, and reservoirs
Service delivery	Managing operational tasks like water billing, meter reading, and system maintenance to improve efficiency
Technology innovation	Introducing advanced tools such as smart meters, renewable energy systems, and real-time water quality monitoring
Capacity building	Providing technical training and knowledge transfer to enhance public sector water management capabilities
Investment and financing	Bridging funding gaps for large-scale water projects through direct investments and performance-based contracts
Policy and regulation support	Collaborating with the government to develop frameworks that balance profitability and equitable water access

In terms of service delivery, private companies can improve the quality and consistency of water provision in urban areas. Outsourcing operational functions such as water billing, meter reading, and maintenance to private operators helps to improve efficiency and customer satisfaction. By adopting innovative management practices, private firms contribute to more reliable water services, addressing some of the challenges faced by resource-constrained public utilities. However, the success of such initiatives depends on the existence of conducive legal and institutional frameworks, which are currently lacking in Zimbabwe (Mutandwa & Zinyama, 2015).

The private sector also introduces innovative technologies that significantly improve urban water management. Tools such as smart meters, water quality monitoring systems, and renewable energy solutions for water treatment and pumping enhance the efficiency and sustainability of water systems. These innovations help reduce water losses, ensure compliance with quality standards, and promote environmentally friendly practices. For example, the adoption of smart water management systems has been identified as a potential area for private sector investment in Zimbabwe's water infrastructure (Vyas-Doorgapersad & Mutandwa, 2023).

Financing and investment from the private sector are instrumental in bridging funding gaps for large-scale water projects. Through public–private partnerships, private investors support projects that public institutions would struggle to finance independently. These investments often come with performance-based contracts that ensure accountability and efficiency in service delivery. Additionally, private sector partnerships facilitate capacity building and knowledge transfer, providing training and technical assistance to enhance the capabilities of public water management authorities. However, attracting private investment into Zimbabwe's water and sanitation sector requires policy reforms that facilitate effective financial intermediation systems and enhance market liquidity (Moyo, 2013).

Despite these benefits, private sector involvement in water governance also presents challenges that must be carefully managed. One key concern is affordability and access, as privatization can lead to higher water tariffs, potentially excluding low-income households. Balancing the profit motives of private entities with the need for equitable water access remains a critical challenge. Effective regulatory oversight is essential to ensure that private sector activities align with public interests. Weak enforcement mechanisms can result in exploitation, substandard services, or environmental harm. Furthermore, the profit-driven nature of the private sector may prioritize short-term gains over long-term sustainability, potentially undermining efforts to conserve water resources and protect ecosystems (Mutandwa & Zinyama, 2015).

To address these challenges, Zimbabwe needs to adopt strategies that enhance the contributions of the private sector while safeguarding public interests. Strengthening regulatory frameworks is crucial to ensure that private operators operate transparently and align their activities with the national development goals. Inclusive partnerships should be promoted, incorporating mechanisms such as subsidies or tiered pricing models to protect vulnerable populations and ensure affordability. The government can also incentivize private sector innovation through tax breaks, grants, or awards for sustainable and cost-effective water management solutions. Robust monitoring systems and public reporting mechanisms should be established to enhance accountability and ensure high-quality service delivery.

4.3.9 Residents' Associations

Residents' associations play a crucial role in urban water governance in Zimbabwe by serving as intermediaries between communities and governing authorities. These associations are typically grassroots organizations formed by residents to advocate for their rights and interests in local governance, including water service delivery (Mapuva, 2011). Their involvement is especially critical in urban areas where issues such as water scarcity, infrastructure deterioration, and inequitable access persist. One of the primary functions of residents' associations is to represent community concerns regarding water management to local authorities and water service providers (Musekiwa & Chatiza, 2015). They act as a collective voice advocating for equitable access, affordable tariffs, and improved service delivery. This is particularly important in marginalized neighbourhoods, where residents often lack direct channels to express their grievances or influence decision-making processes (Mavupa 2020). By consolidating individual concerns into organized advocacy efforts, residents' associations amplify the voices of vulnerable groups and ensure their inclusion in water governance discussions. Table 4.6 shows the roles and responsibilities of residents' associations in urban water governance in Zimbabwe.

Residents' associations also play a key role in monitoring water service delivery and holding providers accountable. They often collect data on water access, quality, and reliability within their communities, which can be used to identify service gaps and inefficiencies. Documenting these issues and presenting evidence-based reports to authorities and associations contribute to transparency, pushing for the necessary

Table 4.6 Roles and responsibilities of residents' associations in urban water governance in Zimbabwe

Role	Responsibilities
Community advocacy	Representing community concerns on water access, tariffs, and service delivery to local authorities and service providers
Monitoring and accountability	Collecting data on water access, quality, and reliability, and holding providers accountable for service gaps
Legal advocacy	Challenging unfair practices or policies that negatively affect residents' water access
Capacity building	Organizing workshops and campaigns to educate communities on water conservation, sustainable use, and infrastructure maintenance
Conflict resolution	Mediating disputes over water allocation and access within communities
Stakeholder collaboration	Partnering with NGOs, the private sector, and academic institutions for technical expertise, funding, and innovative solutions
Policy engagement	Engaging in policy discussions to ensure community needs are reflected in water governance frameworks

improvements in water management systems (Musekiwa & Chatiza, 2015). In some cases, they also engage in legal advocacy, challenging unfair practices or policies that adversely affect residents' access to water (Bhanye, 2023). For instance, the Combined Harare Residents Association has been instrumental in promoting citizen participation in local governance and demanding proper service delivery and accountability from the Harare City Council.

Beyond advocacy, residents' associations frequently engage in capacity-building and community education initiatives. They organize workshops, meetings, and campaigns to raise awareness about water conservation, sustainable usage, and the importance of maintaining local water infrastructure. These activities not only empower residents with knowledge but also foster a sense of collective responsibility for managing shared water resources (Kusena et al., 2017). In informal settlements, where water governance is often fragmented, residents' associations serve as a vital link, helping communities organize themselves to negotiate better services or implement localized water solutions.

Residents' associations also collaborate with other stakeholders, such as NGOs, private sector actors, and academic institutions, to enhance their effectiveness. These partnerships often bring technical expertise, funding, and innovative solutions to address complex water governance challenges. For example, associations may work with NGOs to rehabilitate community boreholes or establish water committees to oversee the maintenance of localized water points.

Despite their contributions, residents' associations face several challenges. Limited financial resources and technical capacity often constrain their ability to operate effectively (Kusena et al., 2017; Musekiwa & Chatiza, 2015). They may also encounter resistance or lack of cooperation from local authorities or water service providers, particularly in cases where their advocacy highlights systemic failures or corruption. Political interference is another concern, as some associations risk being co-opted by political actors, undermining their autonomy and focus on community interests.

To strengthen the role of residents' associations in urban water governance, it is essential to provide them with institutional support and capacity-building opportunities. Training programs on water governance, legal frameworks, and advocacy strategies can enhance their effectiveness. Establishing formal platforms for dialogue between residents' associations and water management authorities can also improve cooperation and ensure that community voices are integrated into policy and decision-making processes.

4.3.10 Water Mafias in Urban Townships

In economically marginalized townships, informal power structures often emerge to fill gaps left by inadequate state-provided water services. One such phenomenon is the rise of *water mafias*, groups or individuals who control access to water resources, often exploiting residents' dependence on boreholes and other alternative water sources. These entities, sometimes claiming to be linked to powerful political parties such as ZANU–PF, have significant implications for urban water governance, community equity, and social justice.

Water mafias are becoming prevalent in townships where municipal water services have either deteriorated or completely failed due to infrastructural decay, corruption, and economic decline. Boreholes, often installed as emergency interventions by NGOs or international donors, have become critical sources of water in these areas. However, the management and operation of these boreholes have, in some cases, been appropriated by informal groups who impose unauthorized control over their use. These groups, colloquially referred to as *water mafias*, often charge exorbitant fees for water access, creating significant financial burdens for residents, particularly in impoverished neighbourhoods. Reports indicate that women collecting water from mafia-controlled boreholes often face harassment and financial extortion in some communities in cities like Harare and Bulawayo. In many cases, these operations are coordinated by politically affiliated individuals, including youth wings of the ruling ZANU–PF party. Leveraging their political connections, these groups secure de facto control over water points, using intimidation and coercion to exclude alternative management structures or competing interests.

A notable example occurred in Bikita East, where members affiliated with ZANU–PF and the Forever Associates of Zimbabwe vandalized newly drilled solar-powered boreholes. These boreholes, funded by opposition party members to alleviate local water shortages, were sabotaged by having large stones and cement blocks placed inside them, rendering them unusable. Community members suspected that the vandals might have even poisoned the water sources, prompting calls for water quality testing (Change Radio, 2023).

In another incident, after losing the August 2023 elections, a ZANU–PF parliamentary candidate for Chinhoyi Constituency, Thomas Chidzomba, dismantled a borehole he had previously installed for the community. This borehole was intended to provide clean and safe water to the residents of Chinhoyi. Following his electoral defeat, Chidzomba initially locked the borehole but later removed the equipment entirely, depriving the community of a vital water source (ZimEye, 2023). Table 4.7 shows the roles, activities, and impacts of water mafias in the urban townships of Zimbabwe.

The actions of the water mafias highlight the politicization of water resources in Zimbabwe, where access to essential services is manipulated for political gain. Such practices not only undermine community trust but also exacerbate existing inequalities, leaving vulnerable populations at the mercy of politically motivated actors. Vulnerable groups, including women, children, and the elderly, are disproportionately affected as they may lack the financial means or social capital to negotiate access to these water points. The exploitation often involves not only financial extortion but also instances of gender-based violence and harassment, with women and girls frequently targeted as they are traditionally responsible for collecting water in many communities. Moreover, the politicization of water access further entrenches inequities. Residents perceived as opposing ZANU–PF or its affiliates may face exclusion from these vital resources or be forced to pay higher

Table 4.7 Roles, activities and impacts of water mafias in Zimbabwe's urban townships

Role/Activity	Impact
Control of boreholes	Water mafias take unauthorized control over boreholes, charging high fees and exploiting residents
Politicization of water access	Politically affiliated groups manipulate water access to reward allies or punish perceived opponents
Exploitation of vulnerable groups	Women and children face gender-based violence and harassment when negotiating access to water from mafia-controlled sources
Sabotage of community initiatives	Water mafias disrupt efforts of residents to manage boreholes or establish equitable access systems through intimidation and sabotage
Extortion and financial burden	Residents are forced to pay exorbitant fees for water, disproportionately affecting low-income households
Undermining formal governance	Water mafias exploit weak regulatory oversight and governance gaps to operate outside legal frameworks, further eroding trust in authorities
Use of coercion and intimidation	Intimidation tactics are used to consolidate control over water points and exclude alternative management structures or competing interests

fees. This dynamic turns water access into a tool for political control, undermining the principles of equity and justice in urban water governance.

The rise of water mafias reflects deeper failures within Zimbabwe's urban water governance framework. These groups thrive in the absence of effective state oversight, weak enforcement of regulations, and limited community participation in water management. Their operations disrupt formal governance systems, erode trust between residents and authorities, and complicate efforts to implement sustainable water management practices. Additionally, the influence of politically affiliated water mafias undermines community-driven initiatives to manage and maintain boreholes. Efforts by residents' associations or local CSOs to establish equitable water management structures are often met with resistance, intimidation, or outright sabotage by these informal groups.

Addressing the issue of water mafias requires approaches that strengthen institutional frameworks and promote accountability. Empowering local communities to manage boreholes through inclusive and transparent structures can reduce the influence of informal power brokers. Community water committees, with clearly defined roles and responsibilities, can help ensure equitable access and foster collective ownership of water resources. Municipal authorities must enhance their capacity to monitor and regulate water access points, including boreholes. This includes implementing robust mechanisms to prevent unauthorized control and ensuring compliance with water management policies.

Efforts should also be made to depoliticize water governance by promoting nonpartisan management structures and holding politically affiliated groups accountable for exploitative practices. Civil society organizations can play a crucial role in promoting transparency and equity in water governance. Addressing the criminal aspects of water mafia operations requires coordinated action from law enforcement agencies. Investigations into extortion, harassment, and other abuses should be prioritized, and perpetrators should face legal consequences. Educating communities about their water rights and equipping them with the tools to resist exploitation is also essential. Capacity-building programs led by NGOs or CSOs can empower residents to advocate for fair access and challenge informal monopolies.

The phenomenon of water mafias highlights the intersection of governance failures, political exploitation, and socioeconomic vulnerabilities in Zimbabwe's urban water management landscape. While these informal structures may temporarily fill gaps in service delivery, their exploitative practices undermine the principles of equity, justice, and sustainability that should underpin urban water governance. Addressing this issue requires coordinated efforts to strengthen institutional capacity, promote transparency, and empower communities to reclaim control over their water resources. Only through inclusive and accountable governance can Zimbabwe overcome the challenges posed by water mafias and ensure water security for all urban residents.

4.4 Institutional Challenges in Urban Water Governance in Zimbabwe

The governance of urban water in Zimbabwe is fraught with a variety of institutional challenges that stem from gaps in legal frameworks, overlapping responsibilities among governmental bodies, and coordination issues across different levels of government. These challenges complicate the management of water resources and impede the provision of consistent and equitable water services.

4.4.1 Gaps in the Existing Legal and Institutional Frameworks

One of the main challenges in Zimbabwe's urban water governance is the presence of significant gaps within the legal and institutional frameworks. These gaps often arise from outdated policies that do not align with current realities or from laws that lack comprehensive provisions for modern water management practices. For instance, Zimbabwe's Water Act, although revised, still does not fully address issues such as water pollution control, integrated water resource management, and climate change adaptation, which are critical in the context of increasing urbanization and environmental degradation. This inadequacy is particularly concerning given the increasing urbanization and environmental degradation in the country. Moreover, there is often a disconnect between policy formulation and its implementation. While the policies may be robust on paper, practical implementation is hindered by limited financial resources, inadequate technical capacity, and lack of enforcement mechanisms. This disconnect leads to a scenario where water governance policies are not effectively translated into actionable outcomes.

For instance, the Auditor-General's report highlights that many local authorities are to blame for residents' severe water challenges, particularly in urban areas. These local authorities fail to expand water infrastructure, repair broken pumps and meters, replace ageing water distribution networks, and curtail illegal connections and vandalism. Such deficiencies lead to shortages of clean and safe water across many local authorities, increasing the opportunity cost of unpaid care work and leading to the spread of diseases such as cholera and typhoid (Sibindi, 2024). Additionally, the centralization of bulk water supply through ZINWA has been criticized as a poorly conceived political strategy rather than a prudent policy for service delivery. This centralization has contributed to the water crises in cities like Bulawayo, where the lack of an inclusive water governance framework involving citizens and relevant stakeholders has exacerbated the situation (Nkomo, 2021). The water crisis in Bulawayo, characterized by water rationing, cutting of water supplies and challenges of quality of water in places such as Luveve, culminating in a number of deaths, is not a new phenomenon. Box 4.1 is an excerpt from a newspaper article demonstrating in detail the water crisis in Bulawayo—a crisis of leadership and water governance.

Box 4.1: Bulawayo Water Crisis: A Crisis of Leadership and Water Governance

The Bulawayo water crisis is not just a crisis of lack of water but more importantly a crisis of leadership and a governance deficit at local and national levels. The shortage of water has bedevilled the city since its establishment in the 19th century and the fact that Bulawayo is situated in natural region four means availability of water will always be a challenge as rainfall is scarce and erratic in this region. . .

Centralisation of bulk water supply through the Zinwa Act was more of a poorly thought-out political strategy of social and political control rather than a prudent policy to deliver services. The fact that government, through its water flagship, built one dam (Mtshabezi) over 40 years and the BCC [Bulawayo City Council] built five dams is abundant testimony to this. It can be argued that devolving of provision of water sources may cause administrative inefficiencies, but there is little empirical evidence to sustain this argument. The fundamental assumption about local authorities being given the mandate to take over bulk water supply is that town councils, city councils and rural district councils have the requisite capacity to deliver on this mandate. It is argued that this was possible previously because of the broad revenue base which, for example, the BCC had and a more conducive economic environment.

I still argue that centralisation especially of bulk water supply has been one of the most monumental policy failures of government which has contributed to the current water crisis. Centralization epitomises water governance failure by the government of Zimbabwe and exposes policy contradiction given the government's posturing on devolution through its "Devolution and

(continued)

Box 4.1 (continued)

Decentralization Policy" and importantly section 264 of the Constitution. There is need to revisit the Zinwa Act and the Water Act to give expression to a policy framework which prioritises an inclusive policy and regulatory framework within a devolved system of local governance and service delivery. The government has spectacularly let down Bulawayo by persistently and consistently failing to allocate adequate resources to address the city's long-term water needs.

Source: Nkomo (2021)

Addressing institutional challenges requires comprehensive reforms to modernize the legal frameworks governing water resources. This includes updating the Water Act to incorporate provisions for pollution control, integrated resource management, and climate resilience. Furthermore, enhancing the capacity of local authorities through adequate funding, technical training, and the establishment of effective enforcement mechanisms is essential to bridge the gap between policy and practice. Promoting inclusive governance that involves community participation can also ensure that water management strategies are responsive to the needs of urban populations.

4.4.2 *Institutional Overlap and Conflict Between Governmental Bodies*

Institutional overlap occurs when multiple agencies or departments share similar mandates without clear delineation of responsibilities. This is prevalent in Zimbabwe as responsibilities for water governance are distributed across several ministries and agencies, including the Ministry of Environment, Water and Climate, ZINWA, and local municipal councils. Such overlap often leads to jurisdictional conflicts and bureaucratic tussles that delay decision-making processes and project implementations. For example, ZINWA has the mandate to manage all national water resources, while local councils are responsible for water service delivery within their jurisdictions. The overlap in roles can lead to conflicts, particularly in terms of resource allocation and operational authority, which in turn affect the efficiency and effectiveness of water management at the urban level. For instance, in Hwange, the Hwange Local Board sought to take over water management from ZINWA, arguing that, according to the Urban Councils Act, this responsibility should lie with local authorities. This situation has led to tensions and delays in water service delivery improvements (Sibindi, 2024).

Similarly, in Gutu, a prolonged dispute between the Gutu Rural District Council and ZINWA over control of the water supply in the Mpandawana town was only resolved after significant delays, which adversely affected residents' access to water. These conflicts are further complicated by financial disagreements. ZINWA has taken legal action against several councils over the payment of levies and license fees for water infrastructure, highlighting the lack of clear guidelines on financial responsibilities between the authority and local councils (Laiton, 2017). Such institutional overlaps and conflicts lead to delays in decision-making and project implementation, ultimately affecting the efficiency and effectiveness of urban water management in Zimbabwe. Clarifying the roles and responsibilities of each entity is essential to improve coordination and service delivery in the water sector.

4.4.3 Coordination Difficulties at Various Levels: Local, Regional, and National

Coordination difficulties are another significant challenge in the governance of urban water in Zimbabwe. Effective water management requires seamless coordination between different levels of government—from local to national—as well as integration of efforts across regions to address shared water challenges. However, coordination is often impaired by several factors, including communication barriers, different priorities, and a lack of resources. For instance, the water crisis in Bulawayo has been attributed not only to natural factors but also to leadership and governance deficits at both local and national levels. The city's water shortages, exacerbated by erratic rainfall, have been compounded by inadequate infrastructure development and maintenance, reflecting a lack of coordinated long-term planning and investment (Nkomo, 2021). Similarly, the Auditor-General's report highlights that many local authorities are responsible for severe water challenges in urban areas due to poor service delivery and financial mismanagement. The report indicates that existing infrastructure lags in meeting the growing demand for services, with backlogs in constructing new infrastructure and maintaining existing facilities. This situation underscores the need for improved coordination and resource allocation among different government levels to effectively address urban water issues (Sibanda, 2024).

Poor communication channels between different governmental layers can result in inconsistent policies and actions. There is frequently a lack of feedback mechanisms to ensure that local challenges and needs inform national water governance strategies. Different government levels may have varying priorities, which can conflict or compete with one another. For example, while national policies might focus on large-scale infrastructure development, local governments may prioritize immediate water supply issues or community-based water conservation strategies. Inequities in resource distribution often led to disparities in water governance capacities between different regions and localities. Wealthier or more influential regions might receive more attention and resources, leaving poorer urban areas struggling with inadequate water services.

Another primary challenge in inter-institutional coordination is the overlapping responsibilities and jurisdictions among different entities. This overlap can lead to conflicts, redundancies, and gaps in service delivery. For example, the responsibilities for water provisioning and quality monitoring might intersect among ZINWA local councils and the Ministry of Health, leading to confusion and inefficiencies. Furthermore, political influence and corruption can disrupt coordinated efforts, as differing political agendas and corruption scandals undermine trust and cooperation among institutions. Budget constraints and funding allocation also pose significant challenges. Limited financial resources result in competition among different bodies for the inadequate funds available, which can delay or halt crucial water management projects. This situation is exacerbated by economic instability, which affects the overall budget and funding priorities of the government.

To overcome these challenges, there is a need for reforms that streamline roles and responsibilities, enhance communication and coordination mechanisms, and ensure that legal frameworks are adaptable to modern challenges in water governance. Addressing these institutional challenges is crucial for improving the resilience and efficiency of urban water systems in Zimbabwe.

Despite these challenges, there have been successes in inter-institutional coordination, particularly when international donors or NGOs are involved. Partnerships with international agencies have facilitated improvements in water infrastructure, capacity building for local authorities, and the implementation of IWRM practices. Community-based initiatives, where local stakeholders, including residents' associations and CSOs, are involved, have also seen successes. These initiatives often lead to more sustainable and appropriate solutions to water management challenges. For instance, participatory water management projects have been effective in both rural and urban settings, improving water access and conservation efforts.

4.4.4 Infrastructure Issues

In Zimbabwe, the state of urban water infrastructure is a critical concern that has far-reaching implications for public health, economic stability, and social equity. The challenges in this sector are multifaceted, stemming from years of underinvestment, lack of maintenance, and insufficient modernization, all of which are compounded by broader economic difficulties. Much of Zimbabwe's water infrastructure, including pipes, treatment facilities, and pumping stations, is aged and deteriorating. In cities like Harare and Bulawayo, the infrastructure was initially designed to cater to much smaller populations. Harare's water infrastructure, for instance, was originally designed in the 1950s to serve approximately 300,000 residents (See Box 4.2). Today, the city's population has surged to over two million, placing immense pressure on the outdated system and leading to frequent pipe bursts and water losses. As these urban areas have expanded, the existing systems have become increasingly inadequate to meet the demand. The ageing pipes are prone to leaks and bursts, leading to significant water losses and contamination risks. In Harare, it is estimated that over 60% of treated water is lost before it reaches consumers, a stark indicator of the inefficiencies within the system (Toriro, 2024).

Box 4.2: Poor Planning, Illegal Settlements Fuel Harare's Worsening Water Crisis

Like in many parts of the country, perennial water shortages in Harare are endemic. Harare and other outlying areas such as Chitungwiza, Norton, Ruwa and Epworth largely depend on Lake Chivero for their water supplies. The capital city's piped water infrastructure, which was developed in the 1950s, initially catered for an estimated population of 300,000 people. However, Harare's increasing population and demand for the precious liquid has seen Lake Chivero catering for an estimated population of five million people. The water hyacinth weed, which has also invaded Lake Chivero is threatening both the quality and volume of water, posing a big threat to Harare's main water supply. Factors such as public sector corruption, non-maintenance of the old water infrastructure, non-availability of water chemicals and the general economic crisis have also been identified as some of the leading causes of Harare's water crisis …

Source: Gweshe (2023)

The capacity of water treatment facilities in urban areas is another major issue. Many of these facilities have not been upgraded to handle the current volumes of water required, nor to adequately remove contaminants from the water supply. This situation is exacerbated during rainy seasons when runoff overwhelms systems, leading to increased pollution levels in water bodies used for urban water supplies. The Morton Jaffray Water Treatment Plant, a key facility for Harare's water supply, was constructed to serve a much smaller population. As a result of obsolete infrastructure, the city loses a significant amount of water through leakages, exacerbating the water crisis. The perennial water crisis, especially in the Harare metro's local authorities, is complicated as it goes beyond obsolete water infrastructure and balloons the population, climate change, inadequate treatment chemicals, and pervasive government corruption and mismanagement.

The lack of proper treatment has direct implications on public health, as evidenced by recurrent cholera and typhoid outbreaks in urban areas like Harare, which are partly attributed to contaminated water supplies (Bhanye et al., 2023). The effectiveness of water infrastructure is also heavily reliant on a stable power supply, which has been a persistent problem across Zimbabwe due to economic challenges and ageing energy infrastructure. Water treatment plants and pumping stations require constant, reliable electricity to operate efficiently. Frequent power outages disrupt water pumping activities, often leaving residents without water for extended periods. Alternative power sources such as generators are not always available or feasible for long-term use due to high operating costs. Box 4.3 gives details on the water crisis in Harare, the capital city of Zimbabwe.

Box 4.3: Inside Harare's Worsening Water Crisis

The main water sources here are boreholes. But most of them are drying up and residents, especially, have a torrid time fetching water in faraway places while some have resorted to the condemned quarry dams. "We fear for the worst with the cholera outbreak in Harare". Harare has a population of 1,849,600, according to the 2022 census, and the province has 2,487,209 people. According to a 2021 report by ZimFact, Harare's water system was designed in the 1960s and produces approximately 704 mega litres (ML) of water per day against an estimated daily demand of between 800ML to 1 300ML. The perennial water crisis in Harare metro's local authorities has been linked to obsolete water infrastructure, a ballooning population, climate change, inadequate treatment chemicals, and pervasive government corruption and mismanagement. The Morton Jeffrey water purification plant was designed for a maximum population of 300,000 back then, before independence, and obsolete infrastructure has seen the council lose a lot of water through leakages.

"The water supply situation in Harare is dire, especially looking at the level of demand per day versus the daily supply capacity," Kumalo said. "The outbreak of cholera at this moment in Harare confirms that we are indeed in a crisis situation." Harare mayor Ian Makone told Human Rights Watch in 2021 that water woes were due to leakages in the old, dilapidated, and inadequate water distribution network. "More than 40% of pumped-treated water is not delivered due to leakages," said Makone, who was a councilor then. Harare Residents Trust director Precious Shumba said the local authority had failed to provide adequate potable water to its residents. "At least 60% of treated water is being lost through leakages and illegal connections along the water distribution network," Shumba said. "This leaves only 40% of treated water to be supplied to connected households." The rest of the population relies on shallow wells and community boreholes sunk by humanitarian agencies, the church, and the council.

"To worsen the situation, people in Glen View, Glen Norah and Budiriro have the misfortune of having poorly laid-down underground sewerage pipes that are blocked, sand-filled and broken. "The result is that the boreholes in these areas yield water that is contaminated." As a result, our water is highly risky to consume, hence the spread of dysentery, diarrhoea and other water-borne diseases. "The environment is also contaminated. Refuse collection has collapsed across most suburbs." To further complicate the situation, the sewerage treatment plants are dysfunctional and untreated sewer is being discharged into the rivers and streams leading to Lake Chivero. "Going forward, underground water and sewerage infrastructure should be urgently replaced."

Source: Kachiko and Muromo (2023)

Financial limitations severely hamper the ability of local authorities and water management bodies like ZINWA and the local authorities to improve and maintain water infrastructure. Recently, Harare's mayor, Jacob Mafume, highlighted the urgent need for a massive overhaul of the city's piping system, estimating the cost at around US$800 million—an amount beyond the city's current financial capacity (Makopa, 2024). Investment in water infrastructure is capital intensive, and the economic difficulties faced by Zimbabwe have resulted in limited public funds available for such investments. This lack of funding is a critical barrier to undertaking the necessary upgrades and expansions of the water systems to meet the needs of growing urban populations.

Rapid urbanization without corresponding upgrades to infrastructure planning has further complicated the challenges. Many urban areas have seen expansion into zones not originally designed for extensive water and sanitation networks, making it difficult to efficiently extend services (Taruvinga, 2024a Informal settlements, in particular, often lack access to centralized water and sanitation services, complicating efforts to integrate these areas into the urban water management system. This situation forces residents to rely on unsafe water sources, increasing the risk of waterborne diseases. Addressing these infrastructure issues requires significant investment, modernization efforts, and improved maintenance practices to ensure safe and reliable water supply in Zimbabwe's urban areas.

4.4.5 Resource Allocation and Accessibility

Effective and equitable resource allocation and accessibility to water remain significant challenges in urban water governance in Zimbabwe, especially in informal settlements. Institutional policies, often shaped by historical, social, and economic contexts, play a crucial role in determining how water resources are distributed and accessed by different segments of the population. Institutional policies in Zimbabwe have traditionally prioritized water distribution to urban centers and formal residential areas, which are seen as economically productive and politically significant. However, these policies often neglect the peripheral and informal settlements, where the need for basic services, including water, is most acute (Matamanda et al., 2024). The allocation of resources is influenced by a complex interplay of political, economic, and social factors, where areas with greater political influence or economic value tend to receive more attention and investment in infrastructure.

In informal settlements, which often develop without formal planning and recognition from local authorities, the lack of infrastructure investment leads to inadequate water supply systems (Maruta et al., 2023). Residents in these areas frequently rely on alternative sources such as unprotected wells, boreholes, and water vendors, which can be costly and unsafe. Institutional neglect is evident through the absence of regularized water supply services, which exacerbates the vulnerabilities of these communities. The accessibility of clean and safe water in informal settlements is hindered not only by inadequate infrastructure but also by the cost and physical availability of water (Shayamunda, 2024). The high cost of water from informal vendors, who often charge prices far above the subsidized rates of municipal supplies, places a significant burden on the poorest urban dwellers. Moreover, the physical hauling of water from distant sources poses severe hardships, particularly affecting women and children, who are usually responsible for water collection. In areas like Mabvuku, a suburb of Harare, residents have endured up to 30 years without reliable tap water, leading to dependence on unsafe water sources and increasing the risk of waterborne diseases (Nyuke, 2021).

Furthermore, the intermittent water supply in many urban areas due to aged infrastructure and frequent breakdowns forces residents to rely on stored water, which increases the risk of contamination. Informal settlements are disproportionately affected by acute water shortages, which are becoming more common as a result of climate variability and mismanagement. These shortages can lead to severe health risks, as inadequate water supply compromises basic hygiene and increases the incidence of waterborne diseases.

Recognizing these challenges, efforts have been made to reform institutional policies to improve water accessibility. Initiatives such as the creation of community water points and the subsidization of water in poorer areas have been implemented to alleviate some of the disparities in water access in urban areas in Zimbabwe. Additionally, some local authorities have begun partnering with international donors and NGOs to develop water infrastructure directly in informal settlements. However, for these reforms to be effective, they must be part of a broader strategy that includes recognizing the rights of residents in informal settlements, involving them in decision-making processes, and ensuring that water governance is inclusive and equitable. Such a strategy requires a shift in institutional priorities and resources to focus on the needs of the most vulnerable populations.

4.4.6 Corruption and Mismanagement: Impact on Water Governance

Corruption and mismanagement within institutions responsible for water governance significantly impede the equitable distribution and effective management of water resources in urban areas (Matamanda et al., 2021). In Zimbabwe, these issues manifest in various forms, from the siphoning of funds meant for water infrastructure projects to the unequal allocation of water resources, all of which have severe repercussions for urban populations. One of the most direct impacts of corruption in water governance is the misallocation and embezzlement of funds. Resources that are supposed to be dedicated to the maintenance and upgrade of water infrastructure often get diverted into the private pockets of corrupt officials. As a result, numerous water projects fail to reach completion, or if completed, they may be of substandard quality, failing to meet the intended needs of the community. For example, audits and investigations have occasionally revealed that money allocated for new water pipelines or treatment plants was misused, leading to continued struggles with water shortages and contamination. A notable example is the misappropriation of a US$144 million loan from the Export–Import Bank of China, intended for the rehabilitation of Harare's water infrastructure (Matenga, 2021). Investigations revealed that a substantial portion of these funds was diverted for personal use by city officials, including the purchase of luxury vehicles, while the intended water projects remained incomplete. Box 4.4 is a newspaper excerpt on cases of corruption, neglect and plunder leading to the water crisis in the city of Harare.

Box 4.4: Harare's Water Crisis: A Case of Corruption, Neglect and Plunder

The perennial water crisis in Harare is most likely a result of corruption that has seen individuals within the local authority turning away good and well-meaning investors and opting for dubious firms for personal benefit... An investigation by The Standard has revealed that council officials entered into deals with companies they would manipulate for personal gain and in doing so failed to follow correct tender processes... Chisango, suspended acting chamber secretary Charles Kandemiri, suspended acting finance director Stanley Ndemera, and acting Harare water director Mabhena Moyo, among others, allegedly pushed for the engagement of four Chinese companies without following tender processes in the 2018 US$868 million loan from the China Eximbank, without going to tender.

Documents at hand show that a report presented to a joint meeting of the environmental management committee that sat together with the finance committee in October 2019 authorised Chisango to engage the four Chinese companies based on a falsified report he had allegedly prepared and presented. Sino Hydro, China Gezhouba Group Company (CGGC), China Geo-Engineering Company (CGEC) and China Machinery and Engineering Company (CEMC) were awarded the contracts that Town House sources say were improperly processed... As part of the deal, the city contracted Sino Hydro to construct Lyndhurst Sewage Treatment Plant and expand Crowborough Sewage Treatment Plant at a cost of US$238 million, yet according to experts who gave advice to council, the same project could be have been completed for US$120,000...

Another scandal that authorities are pushing to further probe is the US$144 million Chinese loan for the refurbishment of the Morton Jaffray Waterworks that has been abandoned with no significant progress having been made. The project was supposed to have been completed in 2016 according to council projections but underhand dealings, inflation of prices and corruption brought the project to a halt. A visit to Morton Jaffray as part of the investigations revealed that the Chinese who had camped at the site have since left, following

(continued)

Box 4.4 (continued)

withholding by the Chinese government of the remainder of US$72 million. The Chinese government cited corruption. Senior council officials, including then town clerk Tendai Mahachi, former director of water, Christopher Zvobgo, and others were accused of inflating prices for personal gain. They allegedly bought over 25 posh vehicles for themselves. The vehicles included three Land Rover Discovery SUVs and Toyota Hilux trucks. It turned out the vehicles were never used on the project.

Source: Matenga (2021)

Bribery in the allocation of water resources is another significant issue. In some cases, businesses and individuals with more financial power may bribe officials to ensure an uninterrupted water supply to their properties or to gain priority over others, including during water rationing periods. This practice not only exacerbates inequality in water access but also undermines trust in public institutions. Furthermore, nepotism can play a role in hiring practices within water management bodies, where jobs are awarded based on familial or political ties rather than merit. This often leads to inefficiency and incompetence in managing water services, as unqualified individuals may not have the necessary expertise to handle complex water management challenges.

Corruption also leads to regulatory capture, where water management entities may become dominated by vested interests that influence regulations and policies to favor specific groups or industries. This can lead to lax enforcement of pollution controls, improper water resource allocation, and the overlooking of illegal water connections, all of which degrade the quality and availability of water for the general population. The lack of enforcement of existing laws and regulations due to corrupt practices ensures that violations go unpunished and standards for safe and equitable water distribution are not upheld.

The consequences of corruption and mismanagement are starkly visible in public health statistics. Communities that suffer from inadequate water supply and sanitation due to corruption are more prone to outbreaks of waterborne diseases such as cholera and typhoid. Moreover, the failure to properly manage and invest in water infrastructure can lead to catastrophic failures, such as the collapse of dams or severe pollution events, posing direct risks to public safety. Addressing these issues requires a multifaceted approach. Transparency initiatives, such as the publication of financial audits and project details, can help reduce embezzlement and ensure that funds are used appropriately. Strengthening legal frameworks and enforcement mechanisms is crucial to tackle bribery and regulatory capture. Furthermore, fostering a culture of accountability through citizen engagement and whistleblower protection can empower individuals to report corruption and demand better services.

4.5 Reforming Urban Water Governance in Zimbabwe: Building Resilient, Inclusive, and Adaptive Institutional Frameworks

In Zimbabwe, effective urban water governance requires the establishment of institutions that are adaptive, transparent, inclusive, and capable of engaging in cross-scalar governance and partnerships. These institutions should be designed to adapt to changing conditions such as population growth, climate variability, and technological advancements, incorporating mechanisms for continuous learning and adjustment in their frameworks and policies. Transparency and accountability are crucial, necessitating open access to information and the involvement of third-party audits to ensure resources are used efficiently and to prevent corruption.

Inclusion is paramount in the formation of these new institutions. They must engage a broad spectrum of stakeholders, including marginalized and vulnerable communities, to address the diverse needs and rights of the urban population, with a special focus on promoting gender equity and involving groups often excluded from decision-making processes. Additionally, institutions must facilitate integration across different sectors and levels of governance—from local to national—while aligning with international water management standards and practices to avoid functional overlaps and ensure coherent action.

Participants in this new institutional framework should include central and local government units with clearly defined and non-overlapping roles, local communities, NGOs, the private sector, and academic and research institutions. Governments should focus on policy formulation, implementation, and enforcement, whereas local communities and NGOs should monitor and manage resources, employing local knowledge in sustainable water practices. Private sector involvement through public–private partnerships can bring the necessary investments and innovations in water infrastructure and management, while academic institutions provide scientific support and promote sustainable practices.

Technology innovation is also vital. Smart water management systems utilizing the Internet of Things (IoT) and Artificial Intelligence (AI) for real-time monitoring and predictive maintenance, along with water-sensitive urban design that incorporates green infrastructure, are essential for efficient resource management and urban planning. Institutions should also foster hydrocitizenship, encouraging citizens to actively participate in water conservation and management efforts, thus fostering a sense of ownership and responsibility towards water resources.

Furthermore, these institutions must align their goals with relevant SDGs, particularly SDG 6 (Clean Water and Sanitation), SDG 11 (Sustainable Cities and Communities), and SDG 13 (Climate Action). This alignment will ensure safe and affordable drinking water and adequate sanitation for all, contributing to sustainable urban communities and climate action initiatives. Equity and social justice must be central to the governance framework, ensuring equitable access to water resources

Table 4.8 Key components for reforming urban water governance in Zimbabwe

Element	Characteristics	Key participants	Innovations and technologies	Community involvement	Alignment with Sustainable Development Goals (SDGs)
Adaptive and flexible	Designed to adapt to changing conditions such as climate variability and population growth	Government agencies, local communities	Smart water management systems, real-time monitoring	Hydro-citizenship	SDG 6 (Clean Water and Sanitation) SDG 11 (Sustainable Cities and Communities), SDG 13 (Climate Action)
Transparent and accountable	Open access to information and third-party audits to prevent corruption	Non-governmental organizations, academic institutions	Water-sensitive urban design	Community-based management	SDG 6 (Clean Water and Sanitation)
Inclusive and participatory	Engages a broad spectrum of stakeholders, promotes gender equity	Local communities, marginalized groups	Internet of Things and AI technologies for predictive maintenance	Community water committees	SDG 11 (Sustainable Cities and Communities)
Integrated and coordinated	Avoids overlapping functions, aligns with international standards	Private sector, government units	Green infrastructure, rainwater harvesting systems	Involvement in planning and management	SDG 13 (Climate Action)

and addressing disparities in water distribution, especially in informal settlements where access is often severely limited.

New governance frameworks must also establish legal mandates that prioritize human health and the environment, promote fair water pricing systems, and ensure water rights to achieve sustainable, equitable urban water management in Zimbabwe. This comprehensive approach will help overcome existing challenges and pave the way for a water-secure future for all urban residents in the country. Table 4.8 provides a structured overview of how different elements contribute to the reformation of urban water governance in Zimbabwe, highlighting the interconnections between institutional characteristics, key participants, innovative technologies, community involvement, and alignment with the SDGs.

4.6 Innovative Approaches and Recommendations for Urban Water Governance in Zimbabwe

As discussed in this chapter, urban water governance in Zimbabwe faced numerous challenges, including ageing infrastructure, inequitable resource distribution, and institutional inefficiencies. To address these issues, innovative approaches are essential to transform governance frameworks and improve service delivery. Zimbabwe can create more sustainable and responsive water management systems through adopting new institutionalism approaches, cross-scalar governance approaches, fostering stakeholder engagement, leveraging technology, and implementing integrated policy reforms. This section presents key strategies and recommendations, emphasizing decentralized management, performance-based regulation, community participation, private sector involvement, and technology-driven solutions to build resilient and equitable urban water governance (Table 4.9).

4.7 Conclusion

This chapter explored the complexities of urban water governance in Zimbabwe through the lens of institutionalism, providing an analytical framework to understand how rules, norms, and institutional arrangements shape governance outcomes. The findings highlight the critical role of institutional structures, stakeholder involvement, and innovative governance models in addressing the persistent challenges of water management in Zimbabwe's urban centers. The interactions between these diverse urban stakeholders are complex and characterized by collaboration and contention. On the one hand, public–private partnerships can harness both public oversight and private efficiency to improve water infrastructure and service delivery. On the other hand, these partnerships can lead to conflicts over priorities, especially if profit motives overshadow public service considerations. Civil society organizations, including NGOs and residents' associations, often play the role of watchdogs, holding both government and private entities accountable. They push for transparency and public participation in decision-making processes related to water governance. However, effective collaboration is frequently challenged by issues such as mismatched expectations, regulatory challenges and resource allocation. Stakeholders often have varying expectations regarding outcomes, timeframes, and financial returns. Inadequate regulatory frameworks can result in disputes and inefficiencies, particularly where roles and responsibilities are not clearly defined. Disagreements over the allocation of financial and human resources can impede project implementation and service delivery.

The institutional challenges faced by cities like Harare and Bulawayo highlight the importance of clear roles and responsibilities among governing bodies. For instance, the ongoing conflicts between the Harare City Council and ZINWA highlight the detrimental effects of jurisdictional overlaps and political interference

Table 4.9 Innovative approaches and recommendations for urban water governance in Zimbabwe

Category	Approach	Description	Expected outcomes
Adopting new institutionalism approaches	Decentralized water management	Transferring authority and resources from central to local governments or community-based organizations to foster localized decision-making and tailored solutions	Increased responsiveness to local needs, enhanced service delivery efficiency
	Performance-based regulation	Holding water utilities accountable based on benchmarks for water quality, reliability, and customer satisfaction, supported by robust monitoring and evaluation frameworks	Improved efficiency, accountability, and customer-focused service delivery
Stakeholder engagement	Community participation	Involving residents directly in the planning, management, and conservation of local water systems to foster ownership and utilize local knowledge for effective solutions	Empowered communities, improved resource management, and enhanced sustainability
	Private sector involvement	Leveraging private investment and expertise through public–private partnerships while ensuring alignment with public interests via enforceable regulatory frameworks	Increased investment in infrastructure, improved efficiency, and expanded access to water services
Technology and innovation	Smart water management systems	Implementing technologies such as smart meters and real-time monitoring systems to track water usage, detect leaks, and optimize resource allocation	Reduced water losses, enhanced resource efficiency, and improved service reliability
	Water-sensitive urban design	Integrating sustainable water management practices into urban planning to create resilient urban environments that address water conservation and climate change adaptation	Improved urban water resilience, enhanced water conservation, and sustainable urban growth
Policy integration and reforms	Legal and regulatory updates	Revising water laws and regulations to reflect current challenges, promote decentralized governance, and ensure equity in water access	Strengthened institutional frameworks and improved governance capacity
	Cross-sectoral policy integration	Aligning water governance with urban planning, housing, health, and climate policies to create synergies and reduce conflicts in resource management	Cohesive and integrated urban management, improved resource allocation, and long-term sustainability

on water service delivery. This aligns with the principles of New Institutionalism, which emphasize the need for coherence and predictability in institutional arrangements to enhance governance outcomes. In contrast, Bulawayo's relatively autonomous water governance demonstrates the potential for decentralized systems to effectively address local challenges when empowered by strong institutional frameworks.

From an institutionalist theory perspective, the chapter illustrated how urban water governance is influenced by the interplay of formal structures, informal norms, and the broader sociopolitical context. Historical institutionalism explains the path dependency seen in Zimbabwe's water governance, where colonial-era infrastructure and management practices continue to shape present-day challenges. Sociological institutionalism sheds light on the importance of community-driven approaches, showing how local participation can reshape institutional norms and foster trust between stakeholders. New institutionalism highlights the need for transparent, performance-driven governance models to address inefficiencies and enhance service delivery.

Effective water governance requires a multi-scalar approach that balances central authority with local autonomy. Decentralized water management offers a better model for empowering local institutions while maintaining accountability. Performance-based regulations and stakeholder engagement are practical tools for improving service delivery, reducing inefficiencies, and fostering equitable access to water resources. Public–private partnerships and community-driven initiatives further demonstrate the potential of collaborative governance in addressing resource and capacity constraints.

Effective urban water governance in Zimbabwe demands the establishment of institutions that are adaptive, transparent, inclusive, and capable of engaging in cross-scalar governance and partnerships. These institutions must be designed to respond to changing conditions such as population growth, climate variability, and technological advancements. They should incorporate mechanisms for continuous learning and adjustment in their frameworks and policies to remain effective over time. Transparency and accountability are essential, necessitating open access to information and the involvement of third-party audits to ensure efficient resource use and prevent corruption.

Inclusion is critical in forming these new institutions. They must engage a broad spectrum of stakeholders, including marginalized and vulnerable communities, addressing the diverse needs and rights of the urban population with a special focus on promoting gender equity and involving groups often excluded from decision-making processes. Furthermore, these institutions should facilitate integration across different sectors and levels of governance—from local to national—while aligning with international water management standards to avoid functional overlaps and ensure coherent action.

The chapter also highlighted the detrimental effects of water mafias in the urban water governance landscape, illustrating how these entities manipulate water access for economic and political gain, exacerbating inequity and undermining public trust in water management systems. Addressing the challenge posed by water mafias requires robust legal frameworks, stringent enforcement of water rights, and enhanced community involvement to ensure equitable water distribution.

Participants in this new institutional framework should include central and local government units with clearly defined and non-overlapping roles, local communities, NGOs, the private sector, and academic institutions. While governmental bodies should focus on policy formulation, implementation, and enforcement, local communities and NGOs can utilize local knowledge for sustainable water management. The private sector, through public–private partnerships, can bring the necessary investments and innovations in water infrastructure and management, whereas academic institutions provide scientific support and promote sustainable practices.

Technological innovation also plays a crucial role in enhancing urban water governance. Smart water management systems using IoT and AI for real-time monitoring and predictive maintenance, alongside water-sensitive urban design that incorporates green infrastructure, are vital for efficient resource management and urban planning. Institutions should foster hydrocitizenship, encouraging citizens to actively participate in water conservation and management efforts, fostering a sense of ownership and responsibility towards water resources.

Moreover, these institutions should align their goals with the relevant SDGs, particularly SDG 6 (Clean Water and Sanitation), SDG 11 (Sustainable Cities and Communities), and SDG 13 (Climate Action). This alignment will ensure safe and affordable drinking water and adequate sanitation for all, contributing to sustainable urban communities and climate action initiatives. Equity and social justice must be central to the governance framework, ensuring equitable access to water resources and addressing disparities in water distribution, particularly in informal settlements where access is often severely limited.

Looking ahead, the future trends in institutionalism and urban water governance in Zimbabwe are likely to be influenced by several factors. Increased decentralization and community-based management could become more prevalent, driven by the need for more responsive and localized water governance structures. Technological advancements such as smart water management systems are expected to play a crucial role in improving efficiency and monitoring, aiding in the reduction of water loss and enhancing service delivery. Furthermore, as climate change continues to have an impact on water availability and demand, adaptive governance models that can handle emerging challenges, such as increased variability in water supply and extreme weather events, will be critical. The incorporation of climate resilience into urban water management planning will likely become a standard practice, necessitating ongoing adjustments to governance frameworks to ensure they remain robust and flexible.

To build resilient urban water governance in Zimbabwe, institutions must embrace adaptive and inclusive practices. This involves creating coherent institutional arrangements, fostering trust and cooperation among stakeholders, and integrating innovative approaches that respond to the unique challenges of each urban center. Drawing on the principles of institutionalism, future efforts should aim to strengthen the alignment between formal and informal rules, social norms,

Table 4.10 Policy recommendations for enhancing urban water governance in Zimbabwe

Category	Recommendation	Description	Expected outcome
Institutional reform	Legal and regulatory updates	Update water laws to reflect current realities, support decentralized management, and encourage community involvement	Strengthened legal frameworks and improved inclusivity in water governance
	Capacity building	Enhance the skills and resources of local water management bodies through training and clear mandates	Increased efficiency and effectiveness of local water management institutions
	Transparency and accountability	Introduce mechanisms like open budgets, audit reports, and public feedback channels for water management operations	Greater trust in institutions and reduced corruption in water governance
Technology and innovation	Smart water management systems	Invest in technologies like smart meters and real-time monitoring systems to reduce water losses and improve efficiency	Enhanced service delivery and optimized resource allocation
	Water-sensitive urban design	Integrate sustainable water management practices into urban planning to promote resilience and sustainability	Improved urban water resilience and sustainable growth
Policy integration	Cross-sectoral policies	Align water governance with housing, health, and environmental policies to create synergies and avoid resource conflicts	Holistic urban development and efficient resource management
	Climate resilience	Incorporate climate change projections and resilience strategies into water management planning	Better preparedness for water scarcity and flood risks due to climate change

and practical governance outcomes, ensuring sustainable and equitable water management for all. The importance of water as a fundamental resource for health, economic stability, and human dignity cannot be overstated. As such, the pursuit of improved water governance is not merely a technical or administrative endeavor but a moral imperative that requires committed action from all sectors of society. Through collaborative efforts, informed by the lessons of the past and oriented towards the future, Zimbabwe can achieve a water-secure future that benefits all its citizens.

For policymakers, the following specific recommendations can be adopted to enhance the institutional framework and overall effectiveness of water governance (Table 4.10).

Through these innovative approaches and recommendations, Zimbabwe can significantly improve its urban water management, making it more sustainable, equitable, and responsive to the needs of its growing urban populations.

References

Bakker, K. (2013). *Privatizing water: governance failure and the world's urban water crisis*. Cornell University Press.

Bandauko, E. (2015). Civil society organizations (CSOs) in urban planning advocacy: Lessons from Zimbabwe. *Journal of Advocacy, Research and Education, 3*(2), 104–109. https://oaji.net/articles/2016/1704-1456921257.pdf

Bates, R. H. (2005). *Beyond the miracle of the market: The political economy of Agrarian development in Kenya*. Cambridge University Press.

Bhanye, J. I. (2023). 'Emerging forms of spatialised and socialised authority 'among 'tenure-insecure peri-urbanites' in African peri-urban spaces: A review study. *Pan-African Conversations, 1*(2), 26–63. https://doi.org/10.36615/pac.v1i2.2734

Bhanye, J. I., Mangara, F., Matamanda, A. R., & Kachena, L. (2023). *COVID-19 lockdowns and the urban poor in Harare, Zimbabwe*. Springer. https://doi.org/10.1007/978-3-031-41669-9

Brooke, A., & Fenner, R. A. (2023). Improving urban water management and building water supply resilience in the city of Harare, Zimbabwe—A systems view. *Civil Engineering and Environmental Systems, 40*(4), 195–228. https://doi.org/10.1080/10286608.2023.2233910

Change Radio. 2023. Bikita East's community solar power boreholes destroyed: Zanu PF and FAZ Wreak Havoc on CCC's Progress. https://changeradiozimbabwe.com/2023/06/29/bikita-easts-community-solar-power-boreholes-destroyed-zanu-pf-and-faz-wreak-havoc-on-cccs-progress/?utm

Cleaver, F. (1999). Paradoxes of participation: Questioning participatory approaches to development. *Journal of International Development, 11*(4), 597–612. https://courses.washington.edu/pbaf531/Cleaver_ParadoxesParticpiation.pdf

Derman, B., & Manzungu, E. (2016). The complex politics of water and power in Zimbabwe: IWRM in the Catchment Councils of Manyame, Mazowe and Sanyati (1993-2001). *Water Alternatives, 9*(3), 513–530. https://www.water-alternatives.org/index.php/alldoc/articles/vol9/v9issue3/331-a9-3-8/file

Dhoba, L. (2022). Strengthening water, sanitation and hygiene governance: A critical review of Zimbabwe's WASH sector institutional arrangements. *H2Open Journal, 5*(2), 248–263. https://doi.org/10.2166/h2oj.2022.063

DiMaggio, P. J., & Powell, W. W. (1983). The iron cage revisited: Institutional isomorphism and collective rationality in organizational fields. *American Sociological Review, 48*(2), 147–160. https://doi.org/10.2307/2095101

Goswami, K. B., & Bisht, P. S. (2017). The role of water resources in socio-economic development. *International Journal for Research in Applied Science and Engineering Technology, 5*(xii), 1669–1674. https://www.ijraset.com/fileserve.php?FID=12295

Government of Zimbabwe. (2017). *National Water Policy*. Ministry of Environment, Water and Climate.

Gweshe, E. (2023). Poor planning, illegal settlements fuel Harare's worsening water crisis. *The Zimbabwe Independent*. https://www.newsday.co.zw/theindependent/news/article/200009966/poor-planning-illegal-settlements-fuel-harares-worsening-water-crisis?utm

Kachiko, M. L. (2023). Inside Harare's worsening water crisis. *The Standard*. https://www.newsday.co.zw/thestandard/news/article/200018419/inside-harares-worsening-water-crisis?utm

Hove, M., & Tirimboi, A. (2011). Assessment of Harare water service delivery. *Journal of Sustainable Development in Africa, 13*(4), 61–84. http://jsd-africa.com/Jsda/Vol13No4_Summer2011_B/PDF/Assessment%20of%20Harare%20Water%20Service%20Delivery.pdf

Koliba, C. J., Meek, J. W., Zia, A., & Mills, R. W. (2017). *Governance networks in public administration and public policy*. Routledge.

Kusena, W., Beckedahl, H., & Desai, S. (2017). Civil society and residents' coping strategies with water shortages and household food insecurity in Gweru, Zimbabwe. *Journal of Arts and Humanities, 6*(2), 91–103. https://doi.org/10.18533/journal.v6i2.1067

Laiton, C. (2017). Zinwa sues 20 councils over levies. *NewsDay Zimbabwe*, 19 June. https://www.newsday.co.zw/news/article/78266/zinwa-sues-20-councils-over-levies?utm

Li, J., Xia, J., Shapiro, D., & Lin, Z. (2018). Institutional compatibility and the internationalization of Chinese SOEs: The moderating role of home subnational institutions. *Journal of World Business, 53*(5), 641–652. https://doi.org/10.1016/j.jwb.2018.02.002

Lima, S., Brochado, A., & Marques, R. C. (2021). Public–private partnerships in the water sector: A review. *Utilities Policy, 69*, Article 101182. https://doi.org/10.1016/j.jup.2021.101182

Makopa, F. (2024). Ageing water infrastructure: Harare appeals for govt support. *Zimbabwe Independent*, 16 August. https://www.newsday.co.zw/theindependent/local-news/article/200030980/ageing-water-infrastructure-harare-appeals-for-govt-support

Mamombe, V., Kim, W., & Choi, Y.-S. (2017). Rainfall variability over Zimbabwe and its relation to large-scale atmosphere–ocean processes. *International Journal of Climatology, 37*(2), 963–971. https://doi.org/10.1002/joc.4752

Mapedza, E., Manzungu, E., Rosen, T., Ncube, P., & Van Koppen, B. (2016). Decentralised water governance in Zimbabwe: Disorder within order. *Water Resources and Rural Development, 8*, 1–11. https://doi.org/10.1016/j.wrr.2016.05.001

Mapuva, J. (2011). Enhancing local governance through local initiatives: Residents' associations in Zimbabwe. *African Journal of History and Culture, 3*(1), 1–12. https://www.academia.edu/77295055/Enhancing_local_governance_through_local_initiatives_Residents_associations_in_Zimbabwe

March, J. G., & Olsen, J. P. (1989). *Rediscovering institutions: The organizational basis of politics*. The Free Press.

Maruta, E., Matsa, M., Mupepi, O., & Dzawanda, B. (2023). Water safety and associated public health impacts in upcoming and informal urban settlements in Zimbabwe: Case of Amsterdam Park in Harare. *Sustainable Water Resources Management, 9*(5), Article 138. https://doi.org/10.1007/s40899-023-00921-3

Matamanda, A. R., Chirisa, I., & Rammile, S. (2021). Elitist domination and its import: Survey of four decades of perpetuation of inequities in Zimbabwe. *Politikon, 48*(3), 450–467. https://doi.org/10.1080/02589346.2021.1952736

Matamanda, A. R., Mphambukeli, T. N., & Chirisa, I. (2024). Exploring water-gender-health nexus in human settlements in Hopley, Harare. *Cities & Health, 8*(1), 44–58. https://doi.org/10.1080/23748834.2022.2136557

Matenga, M. (2021). Harare's water crisis: A case of corruption, neglect and plunder. *The Herald. Newsday*, 6 June. https://www.newsday.co.zw/thestandard/news/article/14572/harares-water-crisis-a-case-of-corruption-neglect-and-plunder?utm

Mazivarimwe, L. (2020). *Assessing the effectiveness of Zimbabwe's water sector framework in addressing climate change induced impacts: Need for adapting the water sector*. (Doctoral dissertation,. Seoul National University, Republic of Korea.

Mgwele, A., Bhanye, J. I., & Mocwagae, K. S. (2024). Water and sanitation for urban health: A gender perspective on impacts and coping strategies in Mangaung Metropolitan Municipality. In A. R. Matamanda & V. Nel (Eds.), *Sustainable development goals and urban health: Strides, challenges and way forward for poor neighborhoods* (pp. 49–67). Springer.

Moyo, P. (2013). An assessment of private sector participation as a viable alternative for improved urban water provision in Zimbabwe: The case of Harare Municipality (Master's thesis, Stellenbosch University, Stellenbosch, South Africa)

Muchadenyika, D., & Williams, J. J. (2018). Politics, centralization and service delivery in urban Zimbabwe. *Journal of Southern African Studies, 44*(5), 833–853.

Musekiwa, N., & Chatiza, K. (2015). Rise in resident associational life in response to service delivery decline by urban councils in Zimbabwe. *Commonwealth Journal of Local Governance, 16/17*, 120–136. https://doi.org/10.5130/cjlg.v0i0.4489

Musemwa, M. (2021). Urban struggles over water scarcity in Harare. *Dædalus, 150*(4), 27–47. https://doi.org/10.1162/daed_a_01871

Mutandwa, H., & Vyas-Doorgapersad, S. (2023a). Public-private partnerships to improve water infrastructure in Zimbabwe. *Insights into Regional Development, 5*(2), 24–37. https://doi.org/10.9770/ird.2023.5.2(2)

Mutandwa, H., & Vyas-Doorgapersad, S. (2023b). The state of water infrastructure development in Zimbabwe. *African Journal of Governance and Development, 12*(2), Article 41. https://hdl.handle.net/10520/ejc-ajgd_v12_n2_a4

Mutandwa, H., & Zinyama, T. (2015). An analysis of the potential use of public-private partnerships in water infrastructural development in Zimbabwe: The case of Harare City Council. *Journal of Public Administration and Governance, 5*(1), 110–136. https://doi.org/10.5296/jpag.v5i1.7366

Nkomo DO. 2021. Bulawayo water crisis: A crisis of leadership, water governance. *NewsDay*, Zimbabwe, 19 February. https://www.newsday.co.zw/opinion-amp-analysis/article/36783/bulawayo-water-crisis-a-crisis-of-leadership-water-governance

North, D. C. (1990). *Institutions, institutional change and economic performance*. Cambridge University Press.

Nyuke, B. (2021). Residents clock 30 years without tap water. *NewsDay, Zimbabwe Independent*, 1 April. https://www.newsday.co.zw/2021/04/residents-clock-30-years-without-tap-water

Ostrom, E. (1990). *Governing the commons: The evolution of institutions for collective action*. Cambridge University Press.

Peters, B. G. (2022). Institutional theory. In C. Ansell & J. Torfing (Eds.), *Handbook on theories of governance* (pp. 323–335). Edward Elgar Publishing.

Pierson, P. (2004). *Politics in time: History, institutions, and social analysis*. Princeton University Press.

Rammile, S., Bhanye, J. I., Lehobo, T., & Hugo, M. (2024). Building resilient urban infrastructure and innovation for urban public health: A case study of Bloemfontein, South Africa. In A. R. Matamanda & V. Nel (Eds.), *Sustainable development goals and urban health: Strides, challenges and way forward for poor neighborhoods* (pp. 69–96). Springer Nature Switzerland.

Saleth, R. M., & Dinar, A. (2004). *The institutional economics of water: A cross-country analysis of institutions and performance*. Edward Elgar Publishing.

Shayamunda, R. H. (2024). Access to basic water and sanitation infrastructure among women in African informal settlements: Implications for health. *Developments in Environmental Science, 15*, 53–76. https://doi.org/10.1016/B978-0-443-21948-1.00004-2

Shepsle, K. A. (1989). Studying institutions: Some lessons from the rational choice approach. *Journal of Theoretical Politics, 1*(2), 131–147. https://doi.org/10.1177/0951692889001002002

Sibanda, Z. (2024). Auditor-general's report: Most local authorities rotten to the core. *The Zimbabwe Independent*, 5 July. https://www.newsday.co.zw/theindependent/opinion/article/200029209/auditor-generals-report-most-local-authorities-rotten-to-the-core?utm

Sibindi, S. (2024). Hwange wants Zinwa to hand over water reticulation services. *The Zimbabwe Independent*, 31 March. https://www.newsday.co.zw/theindependent/local/article/200025040/hwange-wants-zinwa-to-hand-over-water-reticulation-services?utm

Tang, S. Y. (1992). *Institutions and collective action: Self-governance in irrigation*. ICS Press.

Taruvinga, G. R. (2024a). Decentralisation and water governance in Africa: A Zimbabwean case study. *Politeia, 43*(1), 16. https://doi.org/10.25159/2663-6689/17106

Taruvinga, G. R. (2024b). Zimbabwe water crisis: Decentralisation blocked by political interference. *Mail & Guardian*, 25 October. https://mg.co.za/the-green-guardian/2024-10-25-zimbabwe-water-crisis-decentralisation-blocked-by-political-interference/?utm

Thelen, K. (1999). Historical institutionalism in comparative politics. *Annual Review of Political Science, 2*(1), 369–404. https://doi.org/10.1146/annurev.polisci.2.1.369

Toriro, P. (2024). People-centred initiatives as a response to infrastructure collapse in Harare. In I. Chirisa & A. R. Matamanda (Eds.), *Urban infrastructure in Zimbabwe: Departures, divergences and convergences* (pp. 181–199). Springer.

Van Welie, M. J., & Romijn, H. A. (2018). NGOs fostering transitions towards sustainable urban sanitation in low-income countries: Insights from transition management and development studies. *Environmental Science and Policy*. https://doi.org/10.1016/j.envsci.2017.08.011

Vyas-Doorgapersad, S., & Mutandwa, H. (2023). Benefits and challenges to implement public-private partnerships in water infrastructure development in Zimbabwe. *Journal of Economic Development, Environment and People, 12*(1), 57–71. https://doi.org/10.26458/jedep.v12i1.796

Woodhouse, P., & Muller, M. (2017). Water governance—An historical perspective on current debates. *World Development, 92*, 225–241. https://doi.org/10.1016/j.worlddev.2016.11.014

ZimEye. (2023). Zanu PF Aspiring MP rejected by voters removes borehole. *ZimEye News*, 14 March. https://www.zimeye.net/2023/08/26/zanu-pf-aspiring-mp-rejected-by-voters-removes-borehole/?utm

Chapter 5
Unearthing the Political Economy and Ecology of Urban Water Governance in a Failed State

Abstract This chapter explores the political economy and ecology of urban water governance in Zimbabwe. Specifically, the chapter questions the motives and interests of the governance decisions made in managing urban water resources through the political economy and ecology lens. While acknowledging the governance initiatives adopted by the postcolonial government in water governance, for example, the Water Act and the ZINWA Act, the chapter highlights the vested political and economic interests of the decision-makers at the time. Likewise, the dynamics of urban water distribution in Zimbabwe are still greatly influenced by political inequity, which is tarnished by the heavy politicization of water bodies, despite the adoption of these important pieces of law after 1998. Additionally, an often-neglected dimension in urban water governance is the political ecology that brings into perspective the varying interests associated with environmental contests and conflicts that influence how water resources are allocated and shared among urban areas. While climate change has had an impact on urban water resources, the governance of urban water systems has also exacerbated the water woes, hence the term *failed state*. Understanding politics, economics, and ecology is crucial for this chapter because Zimbabwe's current state of obsolete water infrastructure negatively influences the supply of high-quality water services to the public in urban areas. The chapter suggests that to make sustainable political, economic, and ecological functionalities necessary, there is an urgent need to provide and suggest solutions to resolve these urban water governance difficulties. This chapter contributes to the body of knowledge on the need to develop a better political economy and ecological approach to urban water governance dynamics, which have a strong effect on human livelihoods.

Keywords Potable water · Political · Administration dichotomy · Service delivery · Urban water supply · Water governance

A. R. Matamanda et al., *Urban Water Governance in Postcolonial Zimbabwe*, Local and Urban Governance, https://doi.org/10.1007/978-3-031-96973-7_5

5.1 Introduction

Over a million people in Zimbabwe are vulnerable to adverse climate shocks and the recent drought, followed by violent floods, left over 4.2 million Zimbabweans with a need for food and water assistance. This is more than a quarter of the total population of 16.32 million in 2022. There is no doubt that these shocks will limit or even reverse the human development gains that Zimbabwe has made over the years (United Nations Development Programme [UNDP] 2024). The political economy and ecology of urban water refers to the social, economic, and ecological factors that influence the management and distribution of water resources in urban areas. These factors shape how water is sourced, treated, and distributed, as well as who has access to it and who benefits from it (Mgwele et al., 2024). From a political economy perspective, urban water is often seen as a commodity that can be bought and sold, with profits accruing to those who control its production and distribution (Bakker, 2010; Callahan & Mau, 2024). This can lead to unequal access to water resources, with low-income communities often having less access to clean and reliable water than high-income communities. Additionally, Bakker (2010) and the United Nations (2018) argued that privatization of water utilities can result in increased prices and reduced quality of water services for consumers.

Hence, from an ecological perspective, urban water management does have a significant impact on the natural environment, particularly in water-stressed regions. Similarly, over-extraction of groundwater, pollution from urban runoff, and the destruction of wetlands and other natural water sources can all harm ecosystems and reduce the availability of water resources for both human and non-human use (Cooley & Gleick, 2010). Therefore, to address these challenges, urban water management policies and practices must strive to balance the competing demands of economic growth, social equity, and ecological sustainability. All of this requires collaboration and coordination between governments, water utilities, and community stakeholders to ensure that urban water resources are managed in a way that is fair, sustainable, and meets the needs of all urban residents.

The political economy and ecology of urban water in Zimbabwe have been shaped by a complex set of factors that include historical legacies, political instability, economic crisis, environmental degradation, and institutional weaknesses. Zimbabwe has a colonial legacy that has shaped the political economy and ecology of urban water. Kusena (2016) argued how water resources were developed and managed during the colonial period primarily for the benefit of white settlers, while most of the population had limited access to potable water. This legacy has continued to influence the governance of urban water resources in postcolonial Zimbabwe. In the same light, Zimbabwe has experienced political instability and authoritarian rule since independence in 1980. This has had significant impacts on the governance of urban water, consisting of a lack of transparency, accountability, and effective coordination, which have undermined efforts to improve access to safe water (Mapedza et al., 2016).

The UNDP (2024) confirms how Zimbabwe has experienced a prolonged economic crisis since the late 1990s, which has been characterized by hyperinflation, currency devaluation, and high unemployment. This economic crisis has had significant impacts on the urban water sector, including a lack of investment in infrastructure and maintenance, resulting in deteriorating water quality and inadequate service delivery (Bhanye et al., 2023). Furthermore, Zimbabwe has also experienced significant environmental degradation, including deforestation, land degradation, and water pollution, all of which have affected the quality and availability of urban water resources. Furthermore, Zimbabwe's water sector is also characterized by institutional weaknesses, which include a lack of clear policies, inadequate regulations, and a weak institutional capacity (Nhapi, 2009). This has limited the effectiveness of efforts to address the challenges of urban water governance. Hence, this study aimed to unearth the political economy and ecology of urban water governance in a failed state. Based on this aim, the chapter explores the following key research objectives:

- To analyze how political and economic interests influence urban water governance decisions in Zimbabwe.
- To examine the extent to which political inequity shapes disparities in urban water distribution.
- To explore the role of environmental conflicts in exacerbating urban water governance challenges.
- To recommend sustainable governance reforms that integrate political, economic, and ecological considerations for urban water management in Zimbabwe.

5.2 Background

The dynamics of urban water distribution in Zimbabwe have arguably been shaped by a range of social, economic, and political factors, ranging from population growth, corruption, poor governance, overurbanization, infrastructure development, and water resource availability. Despite the significant improvements in water access in recent years, many residents of urban areas in Zimbabwe still lack reliable and affordable access to potable water. Currently, Harare's greater metropolitan area has about 4.5 million people, more than half of whom have no access to clean water and are at risk of waterborne diseases such as cholera and typhoid (UNDP, 2024; United Nations, 2018). All of this is bearing in mind that under section 77 of the Constitution of Zimbabwe, "*every person has the right to safe, clean, and potable water.*" The Constitution of Zimbabwe essentially implies that the government is obligated to take reasonable legislative and other measures, which ensure the accessibility, availability and affordability of potable water to all (Government of Zimbabwe, 2013). However, low-income areas are still underserved, and informal settlements and peri-urban areas have limited or no access to formal water infrastructure. This has resulted in the proliferation of informal water providers, such as borehole owners and water vendors, who often charge high prices for water (Mangizvo, 2014). Mangizvo (2014) further argued that the infrastructure for piped water in Harare was developed in the 1950s, before Zimbabwe's independence in 1980, and designed for a population of 300,000 people. This implies that many water treatment plants and distribution networks were built in the colonial era and have not been adequately maintained or upgraded. This has resulted in frequent breakdowns, leaks, and interruptions in the water supply. With the current population, coupled with overurbanization, the water distribution infrastructure in Zimbabwe's urban areas is aging and in need of urgent repair and maintenance to ensure an efficient and effective potable urban water supply.

The chapter aims to comprehensively investigate the intricate nexus of political economy and ecology in urban water governance within the context of a failed state, focusing on Zimbabwe as a case study. By unpacking the complex interplay between political dynamics, economic structures, and ecological factors, the study sought to further deepen an understanding of the challenges and opportunities in managing urban water resources amid state failure and fragility. Through an in-depth literature review and application of a theoretical framework, this chapter elucidates how political decisions, economic inequalities, and ecological constraints all play a crucial role in shaping water governance practices in Zimbabwe's urban areas. Moreover, the significance of this chapter lies in its potential to inform policy interventions and institutional reforms aimed at fostering more equitable and sustainable urban water management systems in contexts characterized by state failure in line with SDG 6.

5.3 Theoretical Framework: Political–Administration Dichotomy

The political-administration dichotomy theory, often associated with Woodrow Wilson, posits a separation between politics and administration, suggesting that administrative decisions should be made impartially, based on expertise and professional judgment, rather than being influenced by political considerations (Aberbach & Rockman, 2006; Conca, 2006). Thornhill (2014) also confirmed that Wilson agreed with Bluntschli's view that "*administration is not within the scope of politics; the issue of administrative management is not a political issue*". Hence, the political−administration theory has significant implications for urban water governance, particularly in ensuring the efficiency and effectiveness of water management systems. In practice, the theory suggests that while political authorities set the overall goals and priorities for water governance, administrative agencies are also responsible for implementing policies and managing day-to-day operations. This separation helps to insulate technical decisions about water infrastructure, resource allocation, and regulatory enforcement from the volatility of political agendas, promoting stability and continuity in urban water management.

Despite the significance of the theory, unfortunately, the application of the political–administration dichotomy theory in urban water governance is not without challenges. This is because the dynamics surrounding water often intersect with complex political, social, and economic factors, making it difficult to entirely separate administrative decisions from political influence (Conca, 2006). Thornhill (2014) also argued that the accountability of administrative agencies to elected officials can also blur the line between politics and administration, which potentially undermines the impartiality and expertise that the theory aims to uphold. Thus, while the political–administration dichotomy offers a theoretical framework for organizing urban water governance, its implementation in practice still requires careful balancing of administrative autonomy with political oversight to ensure efficiency, effectiveness and democratic accountability in decision-making processes.

Likewise, the effectiveness of the political–administration dichotomy theory in urban water governance also depends on contextual factors such as institutional arrangements, stakeholder dynamics, and the availability of resources. Nhapi (2009) argued that, in some cases, decentralized decision-making structures or public–private partnerships may complicate the clear delineation between politics and administration, necessitating flexible approaches to governance that accommodate diverse interests and objectives. Additionally, Wiek and Larson (2012) asserted how the increasing recognition of the interconnectedness of water systems with broader environmental and social concerns calls for integrated approaches to governance that transcend beyond traditional bureaucratic boundaries. Thus, in the context of this study, while the political–administration dichotomy theory does provide a useful framework for understanding the division of roles and responsibilities in urban water governance, its application must be adapted to local contexts and evolving challenges to ensure the sustainable management of water resources in failed states. This implies that in a failed state, the breakdown of institutional systems can result in a lack of accountability, transparency, and effective coordination, exacerbating the various challenges of water governance.

5.4 Political Economy and Ecology of Urban Water Governance

The political economy and ecology of urban water resources refer to the social, economic, and ecological factors that influence the management and distribution of water resources in urban areas. Although water supply must always be available in sufficient amounts for various purposes, this has not always been the case in several countries, particularly developing countries such as Zimbabwe (Mapedza & Maxwell, 2017). As the World Water Council (2000) noted, the worldwide water crisis is caused not only by a lack of water to meet people's needs but also by a failure to effectively govern water. Indeed, several scholars considered the global water crisis to be primarily a governance crisis (Gandy, 2006; Maramura, 2018;

Swatuk and Kgomotso, 2007). In the same manner, the United Nations (2018) and the World Bank have also long established a direct correlation between countries with severe water shortages and those with poor governance, and such a crisis of urban water governance is evident in Zimbabwe.

Zimbabwe, once known as *the breadbasket of Africa*, has been plagued by many social, economic and political challenges that have significantly affected its urban water governance (Moyo, 2013). The political economy of urban water governance in Zimbabwe is grounded in the history of colonial legacy, post-independence power struggles and neoliberal economic policies (Derman & Manzungu, 2016; Mapedza & Geheb, 2010). Importantly, the ruling party ZANU–PF's control over state institutions has invariably led to patronage and polarization in the allocation of water, thereby deepening the inequalities in access (Matamanda et al., 2024). All these factors have resultantly caused an erosion of public trust in governance institutions owing to mismanagement and corruption. At the same time, the neoliberal agenda, with its emphasis on market-driven reforms and privatization, has been unable to address the water sector's structural inequalities and inefficiencies (Mawere 2018).

Zimbabwe's ecological landscape, characterized by deforestation, land degradation and recurrent droughts, affects urban water governance (Chikozho, 2018). However, while climate change compounds water challenges, Musemwa (2010) contended that Harare's water shortages are not a result of climate change but a multifaceted history of water politics. In other words, although environmental shocks do amplify the vulnerabilities, they are not at the heart of the water crisis. Since independence in 1980, the Government of Zimbabwe has initiated key water sector reform programs. As Moyo (2015) noted, the reforms were aimed at aligning Zimbabwe's water legislation with goals meant to redress inequitable access to water resources. As stipulated in the 1976 Water Act, the reforms were meant to be in sync with the IWRM and also the Zimbabwe National Water Act of 1998. Since Zimbabwe gained independence from British colonial rule in 1980, the country also adopted various water governance initiatives aimed at improving water access, management, and conservation. These initiatives are evident in the Water Act (Chapter 20:24), which is the primary legislation governing water resources management in Zimbabwe. It provides a legal framework for the allocation, use, and protection of water resources, and established the ZINWA as the regulator responsible for managing water resources. The Act was first passed in 1998 and has been amended several times since then to reflect changing priorities and circumstances.

There is also the Zimbabwe National Water Policy (Government of Zimbabwe, 2013/2019), which provides a strategic framework for the sustainable development and management of water resources in Zimbabwe. The National Water Policy emphasizes the importance of community participation, stakeholder engagement, and integrated water resources management in ensuring equitable and sustainable water access for all. The water policy also acknowledges the impacts of climate change on water resources and calls for adaptive and resilient water management strategies (Government of Zimbabwe, 2013/2019). Moreover, the Act seeks to put in place a more egalitarian system of water distribution. Nhapi (2009) argued that this was important in the context of ninety years of colonial rule where water allocation was based on race. While the postcolonial government inherited a well-functioning water supply system, the existing infrastructure was meant to mainly serve European settlers; Africans who lived in segregated and overcrowded townships were only supplied with limited water and sanitation services (Musemwa, 2010). Given these realities, water reforms were a welcome development towards ensuring the equitable and sustainable distribution of water to all. Musemwa (2021:28) argued that, despite all these numerous commitments to improve access to water and sanitation, more than 40 years after independence, the "*inherited unequal distribution patterns of water and sanitation remain deeply entrenched among residents of Harare*".

Shayamunda et al. (2023) affirmed that Zimbabwe's urban water governance failures were laid bare during 2008 and 2009 when a cholera outbreak infected 100,000 people and killed 4300 people, mostly in the City of Harare. Despite the problems with urban water governance, some scholars also underlined the impact consumers have on the performance of water utilities (Maramura, 2018; Musemwa, 2010). In this view, low tariffs and revenue collection can lead to consumers using water inefficiently, thereby leading to high usage and system failures that lead to high maintenance costs. A deterioration of service delivery ensues, which makes consumers unwilling to pay. Once this happens, councils chiefly rely on state subsidies, which often do not materialize. Finally, councils will be unable to pay for wages and other costs, which further compounds the problem, leaving the utilities with a hefty rehabilitation cost (Maramura, 2018). While this view is plausible, adding finances to the water sector without transforming the broader governance problems is unlikely to address systemic urban water challenges.

On the other hand, as the United Nations (2018) noted, addressing governance is crucial in addressing water challenges. In the context of Zimbabwe, urbanization and economic challenges have negatively affected the ability of local authorities to adequately provide water and sanitation. Given this failure, the national government continued to provide grants to assist local water utilities. However, around 2000, the supply of chemicals to treat water became political. The national government, controlled by ZANU–PF, opted for suppliers who were largely aligned with the ruling party, and the MDC mayors and councilors likewise preferred suppliers linked to the MDC or those who were apolitical.

5.5 Politics of Centralization

In Africa, the prominence of oppositional political parties in urban areas has resulted in a dynamic of key politics and service delivery (Bhanye et al., 2023; Kaika, 2003). Given this enlightenment, service delivery is often affected by political contestations between political parties. Examples of countries where this important dynamic is evident include Uganda, South Africa and Zimbabwe (Le Billon, 2013). Liberation parties in many African countries, whose once-popular support seems to be waning, continue to deploy a plethora of strategies to remain in power. One such strategy is meant to frustrate the opposition's success in the areas where they govern, particularly urban areas, in the fear that such success may serve as important stepping stones to win control of the national government as argued by Shayamunda et al. (2023). Although the drive for centralization can be traced back to Zimbabwe's independence in 1980, such politics of centralization became prominent around 2000 when the MDC, the main opposition at the time, emerged as the dominant governing party in urban areas. Within this context, contradictions and tensions over the administration and control of urban areas characterized the relationship between the national government, under the control of the ruling party, ZANU–PF, and local authorities controlled by the Movement for Democratic Change (MDC). ZANU–PF thus sought to regain control of urban areas through the centralization of water.

In the nineteenth century, when major towns were established in the then Rhodesia, urban local authorities had the responsibility to provide water and sanitation services to people. Even after independence, this responsibility rested upon the shoulders of local urban authorities. This ended with the directive of 9 May 2005, which ordered all urban local authorities to surrender water and sanitation functions to ZINWA. Muchadenyika and Williams (2018) argued that the fight to control urban areas has revolved around the centralization of services, such as water and sanitation.

In the years that followed the 2005 directive, as Musemwa (2010:187) noted, the national government "continued to treat ZINWA as an extension of itself, financially propping it up, and defending it to the hilt even in the face of palpable operational deficiencies, to further its political objectives." These political objectives are evident in what Mapedza and Maxwell (2017) concluded to be political meddling in water-related issues in the City of Mutare, and what Manzungu and Mabiza (2004) regarded as attempts by the local government ministry to use water to undermine the MDC-controlled City of Harare. In these contexts, a lack of robust regulatory frameworks and enforcement mechanisms has encouraged a culture of impunity and malfeasance in the urban water sector, and these have undermined transparency and accountability (Muchadenyika and Williams, 2018).

The centralization of water and sanitation services negatively impact on Zimbabwe's towns. Such centralization has tended to marginalize decision-making and service delivery of local authorities at the municipal level (Muchadenyika and Williams). Following the transfer of water responsibilities to ZINWA, the supply of water further deteriorated, with repeated bursts of key water pipelines (Kusena, 2016). Considering the erratic water supply, the United Nations Children's Fund (UNICEF) drilled boreholes in many cities and urban areas. The transfer itself was neither smooth nor well-calculated. Following the transfer of water and sanitation functions from local authorities, ZINWA employed local authorities to bill and collect water charges. However, local authorities sabotaged ZINWA by not collecting water charges properly, thereby reducing revenue for ZINWA; this was corrected when water and sanitation services were transferred back to local authorities (Mawere 2018). This, however, had its challenges as local authorities revised the revenue collection system which led to exorbitant bills that residents were unable to pay. The national government later interfered and ordered local authorities to cancel all outstanding bills. Such interferences emboldened and increased the number of residents who were unwilling to pay for water (Maramura, 2018). This affected the once vibrant tariff collective systems because, since independence, the largest percentage of council revenue came from water, thereby making water an important avenue to raise money for local authorities (Conca, 2006). However, the political and strategic miscalculations affected water and sanitation functions before and after water and sanitation services were handed back to local authorities.

Political interferences in urban governance have handicapped water and sanitation services in Harare (Musemwa, 2010). Water has also been used to punish dissenting voices in areas that criticized the ruling party's rule. Musemwa (2006) aptly captured this in his article titled *Disciplining a 'dissident' city: Hydropolitics in the City of Bulawayo, Matebeleland, Zimbabwe, 1980–1994*. With a particular focus on Bulawayo, Musemwa (2010) highlighted the interplay between ethnic/regional tensions and political disciplining. This harks back memories of the government during the colonial rule when water allocation was segregated between Africans and Europeans in present day Harare. However, Chikozho (2018) extended arguments around centralization to include the national government's desire to access resources in the context of economic decline, the ruling party's ideological belief in centralization to organize the state and to retain power.

Zimbabwe's political economy and ecology for urban water governance is shaped by a complex interplay of historical, socioeconomic and political factors. The move towards centralization is aimed at making the national government relevant in urban areas, where the main opposition often wins (Moyo, 2013). Urban water management is also shaped by governance dynamics which echo an institutional terrain marked by jurisdictional conflicts, overlapping mandates and poor and or lack of coordination (Matamanda et al. 2020). As such, to adequately resolve the crisis of governance in water services is critical in efforts to improve access to water (Walker et al., 2004). This also entails reconfiguring the power dynamics and steer clear from political interferences in water provision.

5.6 Contextualization of a Failed State

Zimbabwe was once celebrated as the *breadbasket of Africa*, but it has spiraled into a state of political turmoil and economic collapse, epitomizing the clear struggles of a failed state (Maramura, 2018). As Mawere (2018) confirms that decades of mismanagement, corruption, and authoritarian rule under Robert Mugabe's regime, followed by a tumultuous transition, have left the country crippled by hyperinflation, widespread poverty, and a crumbling infrastructure. Furthermore, the government's land reform policies, aimed at addressing historical injustices, ultimately led to the collapse of the agricultural sector, which was once the backbone of the economy (Musemwa, 2010). Political repression, human rights abuses, and electoral irregularities further eroded trust in state institutions, contributing to a climate of instability and uncertainty (Moyo, 2013, 2015). International isolation and sanctions also exacerbated the country's plight, with basic services such as healthcare and education in a dire state. Despite sporadic efforts at reform, Zimbabwe continues to grapple with deep-rooted governance challenges, which hinder its prospects for recovery and development.

Zimbabwe's plight extends beyond economic indicators to encompass critical areas like water access and urban governance, further exacerbating its status as a failed state. With an urbanization rate of over 32% as of 2020 (UNDP, 2024), cities like Harare face immense pressure to provide essential services, including clean water. However, water scarcity and inadequate infrastructure plague urban centers, with only around 77% of the urban population having access to improved water sources as of 2017 (United Nations, 2018). Compounded by mismanagement and corruption within local governance structures, the situation worsens. The UNDP (2024) reported that up to 60% of the water supply in urban areas is being lost due to leakages and illegal connections, which is indicative of systemic failures in infrastructure maintenance and governance. Hence, such challenges not only undermine the basic needs of Zimbabweans but also perpetuate a cycle of urban decay, eroding trust in government institutions and exacerbating social tensions.

In Zimbabwe, the concept of political–administration dichotomy faces significant challenges within the context of a failed state. While theoretically advocating for a clear separation between political decision-making and administrative implementation (Aberbach & Rockman, 2006; Conca, 2006), the practical reality of urban water governance in Zimbabwe is marked by pervasive political interference in administrative affairs. This is witnessed with the successive authoritarian regimes which have manipulated state institutions, consequently blurring the lines between politics and administration to consolidate power and suppress dissent. This has resulted in a lack of administrative impartiality and expertise, as key positions are often filled based on political allegiance rather than merit. Both Harvey (2005) and Thornhill (2014) confirmed that the erosion of institutional integrity can exacerbate governance failures, contributing to widespread corruption, economic mismanagement, and the deterioration of public services. Moreover, the politicization of administrative functions has undermined accountability mechanisms, exacerbated social unrest and perpetuated a cycle of authoritarianism and socioeconomic instability in Zimbabwe.

5.7 Discussion and Emerging Issues

There is no doubt that the political–administration dichotomy is an issue in urban water governance in Zimbabwe, as it influences how decisions are made and implemented in the water sector. On the one hand, politics plays an important role in shaping the policies (Svara, 1998) and priorities that guide water governance. Political leaders often have competing demands on their attention and resources and may prioritize short-term gains over longer-term investments in water infrastructure and management (Matamanda et al. 2020; Muchadenyika and Williams 2018). This can lead to uneven or inadequate distribution of resources and a lack of strategic planning for water governance. On the other hand, as highlighted by Chikozho (2018), Conca (2006) and Savenije (2013), the technical aspects of water governance require specialized knowledge and skills and are often best managed by water sector professionals who are trained in engineering, hydrology, and other relevant fields. When political considerations interfere with technical decision-making, it can lead to suboptimal outcomes (Thornhill, 2014) such as underinvestment in water infrastructure, lack of maintenance and repair, and inefficient use of water resources. Addressing the political–administration dichotomy in urban water governance in Zimbabwe will require finding a stable balance between political priorities and technical considerations. This may involve building stronger institutional frameworks for water governance, increasing collaboration between political leaders and water sector professionals, and prioritizing long-term investments in water infrastructure and management.

5.8 Conclusion

This chapter has been explicit in indicating how the water governance system in Zimbabwe has been plagued by an infrastructure deficit that has led to inadequate access to safe and reliable water supplies for many citizens. This infrastructure deficit is rooted in a combination of factors, including underinvestment in the water sector, inadequate maintenance of existing infrastructure, and a lack of coordination between different agencies responsible for water governance. Zimbabwe has struggled to attract investment in the water sector due to a range of factors, including political instability, economic crisis, and a lack of clear policies and regulations. This has limited the government's ability to invest in new infrastructure and maintain existing infrastructure. The failure of government to adequately maintain existing water infrastructure has contributed to the infrastructural deficit. Many water treatment plants, pipelines, and distribution networks are outdated, poorly maintained, or simply non-functional, leading to widespread water shortages and contamination. Zimbabwe's water governance system is highly fragmented, with different agencies responsible for different aspects of water management. This has led to a lack of coordination between agencies and a failure to develop integrated solutions to water governance challenges. In some cases, different agencies have even worked at cross-purposes, exacerbating the infrastructural deficit. In summation, political ecology highlights how power dynamics, governance systems, and environmental factors intersect to shape resource distribution and environmental outcomes, illustrating the need for holistic approaches that address both ecological sustainability and social justice in Zimbabwe's quest for resilience and stability.

The chapter suggests that to make sustainable political, economic, and ecological functionalities necessary, there is an urgent need to provide and suggest solutions to resolving these urban water governance difficulties. This chapter contributes to the body of knowledge on the need to develop a better political economy and ecological approach to urban water governance dynamics since they have a strong effect on human livelihoods. The challenges of water access and urban governance in Zimbabwe underscore the intricate interplay between politics, ecology, and socioeconomic dynamics, forming the essence of political ecology. The mismanagement of water resources and urban infrastructure reflects not only technical failures but also the deeper political and economic structures. Historically, political decisions regarding land redistribution and resource allocation have fueled conflicts over water access and exacerbated ecological degradation. Furthermore, the unequal distribution of water resources often mirrors broader socioeconomic inequalities, with marginalized communities bearing the brunt of water scarcity. Through the application of the political–administration dichotomy, this chapter highlighted the tensions between political influence and administrative governance in Zimbabwe's urban water sector. While governance frameworks such as the Water Act and the ZINWA Act were designed to depoliticize water management, political and economic interests continue to shape decision-making, often at the expense of equitable service delivery. The persistent political inequity in water distribution, coupled with environmental conflicts and deteriorating infrastructure, has further reinforced Zimbabwe's status as a failed state in urban water governance. To address these challenges, there is an urgent need for sustainable governance reforms that separate political interference from technical administration while integrating political, economic, and ecological considerations. Strengthening institutional capacity will also be essential in restoring equitable water governance in Zimbabwe's urban areas.

References

Aberbach, J. D., & Rockman, B. A. (2006). The past and future of political-administrative relations. *International Journal of Public Administration, 29*(12), 977–995. https://doi.org/10.1080/01900690600854589

Bakker, K. J. (2010). Privatizing water, producing scarcity: The Yorkshire drought of 1995. *Economic Geography, 76*(1), 4–27. https://doi.org/10.2307/144538

Bhanye, J. I., Mangara, F., Matamanda, A. R., & Kachena, L. (2023). *COVID-19 lockdowns and the urban poor in Harare*. Springer.

Callahan, R., & Mau, T. A. (2024). Reconceptualizing the politics–administration dichotomy to better understand public leadership in the twenty-first century: A multilateral actors model. *The American Review of Public Administration, 54*(3), 229–241. https://doi.org/10.1177/02750740231213407

Chikozho, C. (2018). Zimbabwe's urban water crisis: Towards a pro-poor policy agenda. *Development Southern Africa, 35*(5), 542–558.

Conca, K. (2006). *Governing water: Contentious transnational politics and global institution building*. MIT Press.

Cooley, H., & Gleick, P. H. (2010). *Water and conflict: Incorporating peacebuilding into water development*. United Nations Development Programme.

Derman, B., & Manzungu, E. (2016). The complex politics of water and power in Zimbabwe: IWRM in the catchment councils of Manyame, Mazowe and Sanyati (1993-2001). *Water Alternatives, 9*(3), 513.

Gandy, M. (2006). *Water, sanitation, and the modern city: colonial and post-colonial experiences in Lagos and Mumbai. Human Development Report 2006*. United Nations Development Programme.

Government of Zimbabwe. (2013) Constitution of Zimbabwe Amendment (No. 20). Government of Zimbabwe: Harare

Government of Zimbabwe. (2019). *Zimbabwe National Water Policy*. Government Printers.

Harvey, D. (2005). *A brief history of neoliberalism*. Oxford University Press.

Kaika, M. (2003). The water framework directive: a new directive for a changing social, political and economic European framework. *Eur Plan Stud, 11*(3), 299–316. https://doi.org/10.1080/0965431030364O

Kusena, W. (2016). *The dynamics of urban water service delivery capacity and the implications for household food security in Gweru, Zimbabwe*. Doctoral dissertation.

Le Billon, P. (2013). *Wars of plunder: conflicts, profits and the politics of resources*. Columbia University Press.

Mangizvo, R. V. (2014). Challenges of water governance in urban Zimbabwe: a case of centralized water governance in the city of Gweru. *Journal of Social Sciences, 41*(3), 353–362. https://doi.org/10.1080/09718923.2014.11893372

Manzungu, E., & Mabiza, C. (2004). Status of water governance in urban areas in Zimbabwe: some preliminary observations from Bulawayo and Harare. *Phys Chem Earth, 29*(15–18), 1167–1172. https://doi.org/10.1016/j.pce.2004.09.003

Mapedza, E., & Geheb, K. (2010). Power dynamics and water reform in the Zimbabwean context: implications for the poor. *Water Policy, 12*(4), 517–527.

Mapedza, E., & Maxwell, A. S. (2017). Water governance in the context of urban water supply in Zimbabwe. In A. S. Maxwell & E. Mapedza (Eds.), *Water governance and collective action: multi-scale challenges* (pp. 95–114). Routledge.

Mapedza, E., Manzungu, E., Rosen, T., Ncube, P., & Van Koppen, B. (2016). Decentralised water governance in Zimbabwe: disorder within order. *Water Resources and Rural Development, 8*, 1–11. https://doi.org/10.1016/j.wrr.2016.05.001

Maramura, T. C. (2018). The efficacy of prepaid water meters for potable water service provision in the Harare City Council, Zimbabwe. (Unpublished doctoral dissertation, Northwest University, Potchefstroom, South Africa)

Matamanda, A. R., Chirisa, I., Dzvimbo, M. A., & Chinozvina, Q. L. (2020). The political economy of Zimbabwean urban informality since 2000–A contemporary governance dilemma. *Development Southern Africa, 37*(4), 694–707.

Matamanda, R., Mphambukeli, T. N., & Chirisa, I. (2024). Exploring water-gender-health nexus in human settlements in Hopley. *Harare Cities & Health, 8*(1), 44–58. https://doi.org/10.1080/23748834.2022.2136557

Mawere (2018) The political economy of poverty, vulnerability and risk management: building bridges of resilience, entrepreneurships and development. Langaa.

Mgwele, A., Bhanye, J. I., & Mocwagae, K. S. (2024). Water and sanitation for urban health: a gender perspective on impacts and coping strategies in Mangaung metropolitan municipality. In A. R. Matamanda & V. Nel (Eds.), *Sustainable development goals and urban health: strides, challenges and way forward for poor neighborhoods* (pp. 49–67). Springer Nature Switzerland.

Moyo, S. (2013). *An assessment of private sector participation as a viable alternative for improved urban water provision in Zimbabwe: the case of Harare municipality*. Master's thesis.

Moyo, S. (2015). Water and power in Zimbabwe: the politics of legitimacy and illegitimacy. *Afr Aff, 114*(454), 246–265.

Muchadenyika, D., & Williams, J. J. (2018). Politics, centralisation and service delivery in urban Zimbabwe. *Journal of Southern African Studies, 44*(5), 833–853.

Musemwa, M. (2021). Urban struggles over water scarcity in Harare. Daedalus, *the Journal of the American Academy of Arts & Sciences, 150*(4), 27–47.

Musemwa, M. (2006). Disciplining a 'dissident' city: Hydropolitics in the city of Bulawayo, Matebeleland, Zimbabwe, 1980-1994. *J South Afr Stud, 32*(2), 239–254. https://doi.org/10.1080/03057070600656119

Musemwa, M. (2010). From 'Sunshine City' to a landscape of disaster: The politics of water, sanitation and disease in Harare, Zimbabwe, 1980–2009. *J Dev Soc, 26*(2), 165–206. https://doi.org/10.1177/0169796X1002600202

Nhapi, I. (2009). The water situation in Harare, Zimbabwe: a policy and management problem. *Water Policy, 11*(2), 221–235. https://doi.org/10.2166/wp.2009.018

Savenije, H. H. G. (2013). Water scarcity indicators: the deception of the numbers. *Physics and Chemistry of the Earth, Part B, 25*(3), 199–204. https://doi.org/10.1016/S1464-1909(00)00004-6

Shayamunda, R., Bhanye, J., Kachena, L., Matamanda, A. R., & Jombo, S. (2023). Gendered experiences of water shortages in Chegutu, Zimbabwe: a participatory action research investigation, 12 December. *Research Square Platform LLC.* https://doi.org/10.21203/rs.3.rs-3731003/v1

Svara, J. H. (1998). The politics-administration dichotomy model as aberration. *Public Adm Rev, 58*(1), 51–58. https://doi.org/10.2307/976889

Swatuk, L. A., & Kgomotso, P. K. (2007). The challenges of supplying water to African cities: reconsidering hydro-political and hydro-social frameworks. *Geoforum, 38*(5), 804–815.

Thornhill, C. (2014). South African municipalities: prospects and challenges: an African perspective. *African Journal of Public Affairs, 7*(4), 140–155. https://repository.up.ac.za/bitstream/handle/2263/45655/Thornhill_South_2014.pdf?se

UNDP (United Nations Development Programme). (2024). *Human development report 2023–24: breaking the gridlock: reimagining cooperation in a polarized world.* United Nations.

United Nations. (2018). *The United Nations world water development report 2018: nature-based solutions for water.* UNESCO.

Walker, B., Holling, C. S., Carpenter, S. R., & Kinzig, A. (2004). Resilience, adaptability and transformability in social-ecological systems. *Ecol Soc, 9(2), Article 5.* http://www.ecologyandsociety.org/vol9/iss2/art5/

Wiek, A., & Larson, K. L. (2012). Water, people, and sustainability—a systems framework for analyzing and assessing water governance regimes. *Water Resour Manag, 26*, 3153–3171.

Chapter 6
Climate Change Adaptation in Urban Water Governance in the Zimbabwean Context

Abstract Climate change is reshaping urban landscapes, demanding transformative approaches to ensure resilience in essential sectors such as water management. Cities need to adapt to remain livable, functional and prosperous in the future. Adaptation is the adjustment process to the actual or expected climate and climate hazards to reduce the negative impacts. Using Zimbabwe as a case study, this chapter examines the adaption of cities to climate change in the context of the urban water sector. The chapter reveals that the impact of climate change compounds complex and multidimensional challenges for secure water service delivery in urban Zimbabwe; from supply shortages to declining groundwater levels. However, due to climate variability, urban residents apply various strategies in adapting to water shortages. These range from reducing water use, using wells and boreholes, using storage tanks, adopting climate-smart agriculture, water reuse and rainwater water harvesting. While policy and institutional reform have advanced at the national level and on paper; nothing much has practically been done from the top to address the visible impacts of climate change on the water sector in urban Zimbabwe. The chapter concludes that adaptation to climate change should be at the center of urban water governance in Zimbabwe. Improved urban water sector resilience and adaptation to climate change can be enhanced through crafting robust policy and ensuring its implementation, good leadership, building strong institutions, community engagement and prioritizing stakeholder collaboration. Emphasizing the need for an integrated approach, the chapter also advocates for the adoption of innovative strategies such as IWRM, green infrastructure, and community-focused approaches like hydrocitizenship to enhance resilience. The chapter calls for a transformative shift in policy and practice, urging policymakers, urban planners, and community leaders to prioritize resilience and actively engage communities in water governance processes.

Keywords Climate change · Resilience · Adaptation · Hydrocitizenship · Urban waterscape · Zimbabwe · Water governance

A. R. Matamanda et al., *Urban Water Governance in Postcolonial Zimbabwe*, Local and Urban Governance, https://doi.org/10.1007/978-3-031-96973-7_6

6.1 Introduction

As the world becomes increasingly urbanized, cities have emerged as critical centers for economic development, social evolution, and environmental impacts (Bhanye et al., 2023; Johnson & Munshi-South, 2017). Today, over 55% of the world's population resides in urban areas, a figure projected to rise to 68% by 2050 (United Nations, 2018). This shift not only concentrates on economic activities and innovations but also exacerbates vulnerabilities to climate change (Ren et al., 2022). Urban areas, particularly in developing countries, face unique challenges in adapting to these changes due to their dense populations and complex infrastructural needs.

The urban water sector plays a pivotal role in maintaining city livability, functionality, and prosperity. Water is the lifeblood of urban ecosystems, supporting everything from domestic needs and sanitation to industry and cooling systems in energy production (Mgwele et al., 2024). However, climate change imposes significant stress on water resources through altered precipitation patterns, increased frequency of extreme weather events, and rising temperatures (Putsoane et al., 2024). These changes threaten the delicate balance needed to maintain urban water systems and necessitate robust adaptation strategies to safeguard continuous access to this critical resource.

In Zimbabwe, the challenges of urban water governance are particularly acute. The country's cities and towns, such as Harare, Bulawayo, Gweru, Chegutu, and Kwekwe, repeatedly face severe water shortages, exacerbated by climate change. These shortages are symptomatic of broader governance issues, including inadequate infrastructure, lack of investment, and management challenges escalating in the face of increasing climate variability (Reniko et al., 2020). For example, Harare's ageing water infrastructure and the siltation of its main water supply, Lake Chivero, significantly reduced the city's capacity to provide for its growing population. Similarly, Bulawayo has experienced periodic water crises that have forced the city to undertake severe rationing measures (Mukuhlani & Nyamupingidza, 2014).

This chapter examines the state of urban water adaptation in the context of climate change in Zimbabwe. It adopts a qualitative methodology, involving a comprehensive analysis of existing literature, policy documents, interviews and case studies relevant to urban water management under changing climatic conditions. The scope of the chapter covers the theoretical frameworks, current governance challenges, and adaptation strategies in practice, and provides recommendations for enhancing resilience in the urban water sector in Zimbabwe. The chapter highlights the critical importance of placing climate change adaptation at the core of urban water governance in Zimbabwe. Examining the interconnected challenges of water scarcity, infrastructure inadequacies, and climate variability, the chapter highlights how current urban water management practices remain insufficient to address the intensifying impacts of climate change. It argues for a paradigm shift that integrates innovative, inclusive, and sustainability-focused approaches, emphasizing community engagement, robust policy frameworks, and adaptive leadership.

The chapter highlights the urgent need to address the mounting risks climate change poses to urban water security, particularly in developing countries like Zimbabwe. Urban water systems are vital to the survival and prosperity of cities, yet they remain highly vulnerable to both environmental and anthropogenic stressors. Providing an in-depth analysis of Zimbabwe's urban water sector, this chapter contributes to the understanding of how climate adaptation strategies can be tailored to the socioeconomic and environmental contexts of resource-constrained settings. It offers insights into practical, locally adaptable solutions for urban water resilience. The chapter not only identifies existing gaps and challenges in urban water adaptation in the context of climate change but also proposes actionable strategies, such as IWRM and hydrocitizenship, that prioritize sustainability and inclusivity in water practices. The integration of community perspectives and case studies enriches the discussion, offering a roadmap for policymakers, urban planners, and community leaders. Advocating for collaborative and transformative approaches, the chapter inspires a shift towards resilient and adaptive water governance systems, with lessons applicable to other cities facing similar challenges worldwide.

6.2 Climate Change and Urban Challenges

Climate change adaptation has gained increasing attention in recent years as the urgent need to respond to the impacts of climate change becomes clearer. In particular, the urban water sector is a critical area requiring adaptation efforts, given the high concentration of people and assets located in urban areas and their dependence on reliable and safe water supplies. Climate change refers to long-term shifts and alterations in temperature, precipitation, wind patterns, and other elements of the Earth's climate system (Adedeji et al., 2014). These changes can occur naturally over geological timescales due to factors such as volcanic eruptions, solar variations, or changes in Earth's orbit (Allan et al., 2023). While climate variability is a natural component of Earth's climate system, the vast majority of recent changes have been driven by human activities, primarily land-use changes, industrialization and the burning of fossil fuels, which increases the concentration of greenhouse gases in the atmosphere (Masson-Delmotte et al., 2021). This anthropogenic influence has led to global warming, characterized by rising global surface temperatures, melting ice caps, rising sea levels, and more frequent and severe weather events such as hurricanes, droughts, and floods (Allan et al., 2023).

Urban areas, by definition, are densely populated regions where the landscape is predominantly built up and covered by impervious surfaces like concrete and tarmac (Weeks, 2010). These areas are centers of commerce, administration, and industry, and they play a pivotal role in the economic development of a country. However, the concentration of population, infrastructure, and economic activities in urban areas makes them particularly vulnerable to the impacts of climate change (Bhanye et al., 2023; Johnson & Munshi-South, 2017). For instance, the urban heat island effect, where urban regions experience higher temperatures than their rural surroundings due to human activities, exacerbates the heat stress experienced during heat waves, increasing energy consumption for cooling and elevating mortality risks (Mohajerani et al., 2017).

Moreover, the impermeable surfaces in urban environments significantly alter natural water drainage patterns, exacerbating flood risks during heavy rainfall events. This situation is further complicated because many cities, especially those in developing countries, are located in coastal areas or by rivers, making them inherently susceptible to sea-level rise and riverine flooding. Recent studies predict significant increases in flood risk for coastal cities due to climate change-induced sea-level rise and storm surges (Hallegatte et al., 2013; Takagi et al., 2016). The challenges posed by climate change are thus particularly pronounced in urban settings, where inadequate infrastructure, governance challenges, and social inequalities often hinder the capacity to manage these risks. These challenges require comprehensive strategies that encompass not only technical and infrastructural adjustments but also socioeconomic and policy-oriented approaches to enhance the resilience of urban systems and communities.

6.3 Heightened Climate Risks in Cities of the Global South

Climate change manifests through a variety of global impacts that transcend regional boundaries, affecting ecosystems and human populations alike. At the most fundamental level, the increased concentration of greenhouse gases has led to a rise in global surface temperatures. The report of the World Meteorological Organization (WMO) confirmed that 2023 was the warmest year on record, with the global average near-surface temperature at 1.45 °C, with a margin of uncertainty of ±0.12 °C above the preindustrial baseline. It was the warmest 10-year period on record (WMO, 2023). This warming contributes to the melting of ice caps and glaciers, leading to rising sea levels that threaten coastal areas with flooding, erosion, and salinization of freshwater bodies (Hallegatte et al., 2013; Takagi et al., 2016). Extreme weather events, such as hurricanes, droughts, and heat waves, have become more frequent and severe due to climate change (Putsoane et al., 2024). These events disrupt food production, degrade water quality, and can trigger large-scale displacements of populations (Diffenbaugh & Burke, 2019). Furthermore, changes in precipitation patterns result in altered hydrological cycles, increasing the likelihood of both severe droughts and flooding, which challenge existing water management systems and exacerbate water scarcity issues globally.

Cities in the Global South are particularly vulnerable to the impacts of climate change, largely due to their rapid urbanization, the high density of the populations, and the significant number of informal settlements. These settlements are often characterized by inadequate housing, lack of basic services, and precarious locations prone to environmental hazards (Rammile et al., 2024; Shayamunda, 2024). For example, informal settlements near riverbanks or coastal zones in cities like Jakarta and Lagos are highly susceptible to flooding and sea-level rise (Revi et al.,

2014). The socioeconomic implications are severe, as many inhabitants of these areas are already disadvantaged by poverty and have limited capacity to respond to climate stress. The urban poor in these contexts often face an accumulation of vulnerabilities, including exposure to extreme heat without adequate shelter or cooling, and increased health risks from vector-borne diseases such as malaria and dengue, which are proliferating in new areas due to changing climate conditions (Zhao et al., 2022). Urban areas in the Global South face specific climate hazards that pose significant risks to infrastructure and water resources (Table 6.1).

6.4 Theoretical Framework

This chapter adopted three primary theoretical approaches to help in understanding the complexities of climate change adaptation and water governance in urban settings. Namely the hydrocitizenship theory, the resilience theory, and adaptive management. In Zimbabwe's urban areas, where water governance faces challenges from aged infrastructure, limited financial resources, and institutional inefficiencies, these theories provide a robust framework for understanding the complexities of climate change adaptation. Each theory complements the others, offering a comprehensive approach that can address both the physical realities of water

Table 6.1 Climate hazards and their impacts on urban infrastructure and water resources

Climate hazard	Description	Implications for urban infrastructure and water resources
Flooding	Increased frequency and intensity of rainfall, coupled with a rise in sea levels, result in frequent and severe urban flooding	Damages infrastructure Disrupts transportation Worsens water quality Increases the risk of waterborne diseases Cities with poor drainage and flood management systems are particularly at risk
Heat waves	Urban heat islands effect and exacerbate the impact of global warming in cities, where temperatures can be significantly higher than in surrounding rural areas	Increases energy demand for cooling Elevates heat-related morbidity and mortality rates
Droughts	Altered precipitation patterns and increased evaporation rates lead to prolonged droughts	Strains urban water supplies Impacts domestic and industrial water use Cities reliant on surface water face challenges in maintaining water supply
Infrastructure stress	Climate change has an impact on the structural integrity of urban infrastructure	Weakening and expansion of materials used in roads and bridges due to increased temperatures Flooding can undermine foundations and damage essential services such as electricity and water supply

management and the social dynamics necessary for sustainable governance. For example, combining hydrocitizenship with resilience and adaptive management could lead to governance systems that are not only physically robust but also socially equitable and adaptive to the changing climate.

6.4.1 Hydrocitizenship Theory

Hydrocitizenship theory explores the relationship between communities and water within the context of democratic engagement and environmental stewardship (Scott-Bottoms & Roe, 2020). It emphasizes the roles and responsibilities of citizens as active participants in water governance, advocating for a more inclusive and collaborative approach to managing water resources (Jones et al., 2021). This theory posits that effective water management is inherently linked to citizen engagement and the fostering of a collective identity centered around water (Kaika, 2004). Hydrocitizenship encourages local communities to partake in decision-making processes, ensuring that water policies are not only sustainable but also equitable. In Zimbabwe, where urban water governance issues are often marked by disparities in water access and quality, particularly in marginalized communities, applying hydrocitizenship could transform water management. Involving urban residents in governance processes, especially in cities like Harare and Bulawayo that are plagued by chronic water shortages and contamination, hydrocitizenship could enhance resilience and ensure that the voices of the most affected populations are heard in the planning and implementation of adaptation strategies.

6.4.2 Resilience Theory

Resilience theory provides a framework for understanding the capacity of systems—be it ecological, social, or infrastructural—to respond to disturbances and still retain their basic function and structure (Hayes et al., 2019). In the context of climate change, this theory is crucial for developing strategies that allow urban water systems to absorb shocks and stresses without catastrophic failure (Folke et al., 2010). Resilience in urban water governance involves creating systems that are flexible, redundant, and capable of self-organization, as well as learning and adapting to new circumstances. The application of the resilience theory in Zimbabwean cities would entail designing water infrastructures that can withstand variable climate conditions and recover quickly from disruptions such as droughts or floods. This includes diversifying water sources, improving infrastructure design, and implementing effective disaster risk management practices. For instance, building water retention facilities can help manage stormwater and mitigate flooding, thereby increasing the system's resilience to climate-related hazards.

6.4.3 Adaptive Management

Adaptive management is a structured, iterative process of robust decision-making in the face of uncertainty, to reduce uncertainty over time via system monitoring (Marmorek et al., 2019). In the context of water governance, adaptive management involves adjusting policies and practices according to systematic monitoring of outcomes and changing environmental conditions (Pahl-Wostl et al., 2007). This approach is ideal for managing complex and dynamic urban water systems where predicting outcomes can be difficult due to the unpredictable nature of climate change. Implementing adaptive management in the urban water sector of Zimbabwe requires a flexible approach to policymaking and infrastructure development. Urban water managers could use real-time data to make informed decisions about water distribution, conservation measures, and emergency responses. For instance, in response to reduced rainfall, adaptive management strategies may include the temporary reallocation of water supplies or the modification of usage restrictions.

6.5 Urban Water Sector in Zimbabwe

The urban water sector in Zimbabwe is structured under the Ministry of Environment, Water and Climate, which oversees water resource management and policy direction. At municipal level, urban councils are responsible for water delivery and sanitation services. However, the actual management of water resources and supply systems often involves multiple stakeholders, including ZINWA that manages the country's major dams and water bodies, and local authorities who handle the distribution of water to urban households (Chikozho, 2010; Manzungu & Mabiza, 2004). Despite the clear delineation of responsibilities, the governance of water resources in Zimbabwe is fraught with challenges related to coordination and efficiency. The dual management model between ZINWA and local authorities has led to conflicts and inconsistencies in policy implementation, adversely affecting water service delivery in urban areas (Zinyama, 2021).

Urban areas in Zimbabwe, particularly Harare and Bulawayo, the largest cities in the country, have experienced chronic water shortages for years. These shortages are largely attributed to inadequate and ageing infrastructure, limited investment in new water sources, and recurrent droughts exacerbated by climate change. The situation is further compounded by significant losses due to water leakages and unaccounted-for water, which in some urban areas amounts to more than 60% of the water supplied (Ndunguru & Hoko, 2016). Many urban residents and businesses have resorted to drilling boreholes in response to persistent water supply shortages. This has led to unregulated groundwater extraction, raising concerns about the sustainability of aquifers and the potential for land subsidence. Groundwater levels in some urban areas have been reported to drop by several meters, indicating significant over-extraction. Another critical issue is that the water infrastructure in many

Zimbabwean cities is outdated and in disrepair. For example, in Harare, some water distribution networks are over 60 years old and suffer from neglect and lack of maintenance (Toriro, 2024). This results in frequent breakdowns and water losses, severely affecting the quality and reliability of water service delivery. The funding for upgrading and maintaining water infrastructure has been inadequate, further exacerbating the problem.

The management of water resources in Zimbabwe is heavily influenced by socio-economic and political factors. Economic challenges, such as hyperinflation and budget deficits, have severely limited the government's ability to invest in necessary infrastructure improvements (Bhanye et al., 2023). Furthermore, political instability and governance issues have led to mismanagement and corruption within the water sector, diverting scarce resources away from essential services (Ndebele-Murisa & Mubaya, 2019). Social factors also play a crucial role. Inequities in water access are prevalent, with low-income and informal settlements often lacking reliable access to water services. This situation exacerbates social tensions and contributes to public health crises, as these communities are forced to rely on unsafe water sources (Zhao et al., 2022).

Zimbabwe's urban water sector faces a complex array of challenges that require comprehensive and integrated solutions. At the national level, Zimbabwe has undergone significant policy reforms to address the impact of climate change on the urban water sector. These reforms aim to create a resilient and adaptive framework that ensures the sustainability of water resources in urban areas. National policies encompass a spectrum of strategies, including climate-resilient infrastructure development, water resource management, and community engagement. These initiatives highlight the recognition of climate change as a critical factor influencing water governance at the highest levels of decision-making. Policy measures and programs aimed at fostering climate resilience and sustainability encompass the following:

- **National Climate Policy**: Establishes foundational guidelines promoting coordinated actions towards low-carbon development and climate-resilient growth pathways.
- **Nationally Determined Contribution**: Outlines ambitious targets to reduce greenhouse gas emissions and fortify adaptive capacities, incorporating specific goals tailored for the water sector.
- **National Adaptation Program of Action**: Articulates priority interventions addressing urgent adaptation imperatives, supported by detailed vulnerability assessments and strategic planning exercises.
- **Zimbabwe Agenda for Sustainable Socio-economic Transformation**: Integrates climate change considerations into broader development blueprints, emphasizing rural and urban water management as vital pillars of socioeconomic well-being.
- **National Water Resources Management Strategy**: Provides a comprehensive roadmap charting desired trajectories for improved water governance, embracing climate change adaptation as a core tenet.

Despite the existence of well-formulated policies, the translation of these policies into effective, on-the-ground implementation faces formidable challenges. Implementation challenges in Zimbabwe's urban water governance include financial constraints, technical capacity limitations, and institutional barriers. The complex nature of climate change adaptation requires a coordinated effort from various stakeholders, and the absence of adequate resources and capacities hinders the timely and efficient execution of planned interventions. A notable gap exists between the formulation of policies and their practical implementation within the urban water sector. While policies may outline comprehensive strategies for climate change adaptation, the effective execution of these strategies often falls short in practice. Factors contributing to this gap include a lack of enforcement mechanisms, insufficient monitoring and evaluation frameworks, and limited community involvement in policy implementation. Bridging this divide is crucial for ensuring that the intended benefits of national-level policies translate into tangible improvements in the resilience of the urban water sector.

6.6 Climate Change Impacts on the Urban Water Sector in Zimbabwe

Zimbabwe, like most African nations, is highly susceptible to the consequences of anthropogenic climate change. According to the IPCC (2021) report, Southern Africa, including Zimbabwe, can expect considerable temperature increases ranging between 1.5 °C and 4 °C by mid-century, relative to preindustrial levels under different emission scenarios. Similarly, mean annual precipitation is projected to decrease significantly in some regions, while others may witness intensified rainfall events leading to flooding incidents (Putsoane et al., 2024). Furthermore, there is heightened uncertainty around future El Niño–Southern Oscillation cycles, which pose substantial ramifications for regional hydroclimatology, given their influence on seasonal rainfall patterns (Perry et al., 2020).

Given the anticipated climate fluctuations, urban water sectors in Zimbabwe confront myriad challenges, notably supply shortages, falling groundwater levels and augmented flood risks. Declining precipitation volumes and prolonged dry spells threaten surface water availability in reservoirs supplying municipal demands. Coupled with growing demographic pressures, dwindling per capita water shares risk compromising essential public services, hygiene standards, and livelihood prospects. Increased temperatures and shifting rainfall regimes adversely affect groundwater recharge rates, resulting in reduced aquifer yields and diminished baseflow contributions to streams and rivers (Kløve et al., 2014). For densely populated urban settlements heavily reliant on borehole abstractions, depleted reserves imply augmented competition for scarce resources, potentially triggering conflicts among users. Intensified convective storms elevate the likelihood of flash floods inundating low-lying residential zones situated along river courses or poorly

Table 6.2 Impacts of climate change on urban water systems in Zimbabwe

Impact category	Description	Specific issues and examples
Water scarcity	Intensification of water scarcity due to climate change	Decreased annual rainfall and increased frequency of droughts leading to reduced water capacity in reservoirs and dams Chronic shortages and competition for water resources among agricultural, industrial, and domestic users
Water quality degradation	The impact of climate change on water quality	Increased temperatures and altered precipitation patterns causing proliferation of waterborne pathogens and pollutants Algal blooms in high heat releasing toxins and depleting oxygen, affecting water quality and aquatic life
Flooding	Increased risk and occurrence of urban flooding	Poor drainage and impervious surfaces exacerbate runoff problems Flash flooding in cities like Harare and Bulawayo damages infrastructure and disrupts water services
Infrastructure stress	Climate change-induced stress on water infrastructure	Increased frequency of extreme weather events causes infrastructure stress Drought conditions reduce water flow, leading to pipe clogs; heavy rains cause infrastructure damage
Health risks	Health issues arising from climate change-related water problems	Water scarcity and poor quality water increase diarrheal diseases Floods provide breeding grounds for disease vectors, increasing the spread of diseases such as malaria and dengue fever
Ecosystem impacts	Effects of climate change on local ecosystems through changes in water resources	Alterations in water availability and quality affect urban and peri-urban ecosystems Reduced water levels and quality in urban wetlands have an impact on wildlife and ecosystem services such as water filtration and flood mitigation

drained catchments (Cutter et al., 2018). Apart from property damage, such occurrences pose severe health concerns stemming from contaminated runoff entering potable water networks or stagnant pools breeding disease vectors (Zhao et al., 2022).

Thus, climate change is a pivotal force reshaping the urban water landscape in Zimbabwe. Both directly and indirectly, climate-induced alterations manifest in various ways in the urbanscape, affecting water quantity, quality, infrastructure, public health, and local ecosystems. The overarching consequences of these changes challenge the capacity of urban areas to provide reliable and safe water services, demanding immediate attention to adaptation and mitigation strategies. Table 6.2 is a summary of the impact of climate change on urban water systems in Zimbabwe.

6.7 Complexity and Multi-Dimensionality of the Climate Change–Water Challenge

As elucidated above, climate change imposes profound implications upon urban water provisioning structures in Zimbabwe. Nevertheless, isolating these influences overlooks the concurrent effect of additional stressors, complicating matters exponentially. Some notable confounders include rapid urbanization, outdated infrastructure, and insufficient funding. Zimbabwe, like many other developing countries, is experiencing rapid urbanization, characterized by the unplanned expansion of built environments. This unchecked growth leads to extensive land-cover modifications, including deforestation and paving over natural surfaces, which disrupt the hydrological cycle. As a result, the natural ability of the land to absorb and store water is compromised, exacerbating water scarcity issues in urban areas. Moreover, the haphazard development often leads to encroachment on wetlands and other critical water resources, further exacerbating the situation.

The urban water infrastructure in Zimbabwe is largely outdated and poorly maintained. Decrepit conveyance channels, treatment plants, and storage facilities suffer from years of neglect and lack of investment. These ageing infrastructure components are not only inefficient but also prone to frequent breakdowns and leaks, resulting in significant water losses. Moreover, incapacity to handle the increasing demand exacerbated by urbanization puts additional strain on the already limited water resources. Upgrading and refurbishing this infrastructure is imperative to improve efficiency, reduce water losses, and ensure a reliable water supply to urban populations.

Limited financial resources pose a significant barrier to addressing the challenges posed by climate change on urban water provisioning in Zimbabwe. Persistent fiscal constraints hinder the government's ability to allocate adequate funds for the maintenance and rehabilitation of water infrastructure. This lack of investment not only perpetuates the deterioration of existing infrastructure but also inhibits the adoption of innovative technologies and practices that could enhance water management and resilience. Additionally, insufficient finances constrain the government's capacity to invest in alternative water sources and diversification strategies, leaving urban areas vulnerable to fluctuations in climate patterns and external perturbations such as droughts or floods.

Addressing these complex challenges requires a multi-pronged approach that integrates urban planning, infrastructure development, financial management, and climate adaptation strategies. It necessitates coordinated efforts from various stakeholders, including government agencies, water utilities, CSOs, and the private sector, to mobilize resources, implement sustainable solutions, and build resilience against the impact of climate change on urban water provisioning in Zimbabwe. Henceforth, comprehending the nexus between climate change and urban water challenges entails acknowledging the convoluted interactions governing these phenomena, demanding integrated solutions capable of surmounting inherent obstacles synergistically rather than sequentially.

6.8 Urban Responses to Climate Change-Induced Water Shortages

Faced with mounting water insecurity, urbanites devise ingenious tactics to circumvent restrictions imposed by fluctuating climatological conditions. These strategies vary in scope and effectiveness, based on the diverse socioeconomic and geographical landscapes across the country. Principal approaches encompass borehole drilling, use of wells, reducing water use, utilizing storage tanks, adopting climate-smart agriculture, water reuse, rainwater harvesting, water rationing and scheduled supply (Fig. 6.1).

Fig. 6.1 Urban responses to climate change-induced water shortages. (Source: Authors)

6.8.1 *Borehole Drilling*

In response to unreliable municipal water supplies and increasing urban water demand, many communities and businesses in Zimbabwe have turned to borehole drilling as a sustainable water solution. Boreholes tap into underground aquifers, providing access to clean and reliable water sources. This strategy has become particularly widespread in urban areas such as Chegutu, Harare, Gweru and Masvingo, where groundwater is relatively accessible and serves as a critical alternative during periods of water rationing or supply disruptions. Borehole drilling has proven to be an effective means of enhancing water security, especially for households, schools, and small businesses that cannot rely solely on municipal supplies. Figure 6.2 shows a community borehole in Gweru.

The benefits of boreholes are manifold. They provide a dependable water source, reducing reliance on inconsistent municipal supplies and alleviating the pressure on shared water resources. This reliability is particularly important for urban residents during prolonged dry spells or in areas where municipal infrastructure is outdated or poorly maintained. Mr Mutasa, a resident of Masvingo, reflected on the impact of borehole drilling in his community:

> *Before we had our borehole, water cuts were a constant struggle. Now, we have enough water for our daily needs, and we even share with our neighbors when their taps run dry.*

Figure 6.3 shows a shared community borehole.

Despite some advantages, the rise in private borehole drilling raises concerns about groundwater management. In some urban centers, the lack of regulation and

Fig. 6.2 Community borehole in Gweru. (Source: Authors)

Fig. 6.3 Shared community borehole. (Source: Authors)

monitoring of borehole drilling activities threatens the sustainability of aquifers. Over-extraction of groundwater can lead to significant depletion, reducing the availability of this resource for future generations. In addition, poor drilling practices and unregulated borehole installations increase the risk of contamination, particularly in densely populated urban areas where wastewater systems may not be adequately managed.

Another challenge is that some community or shared boreholes are very far away from some households. Women and girls are the most affected by this challenge. Given that women are socially prescribed roles that have to do with the use of water at the household level, they are also deemed responsible for ensuring water availability at the household level (Fig. 6.4).

> *For us women and girls fetching water is an industrial job, many households do not have access to wheelbarrows in order to carry more than one container at once. Therefore, they have to carry the water containers on their heads. This is quite tiresome, carrying a 20-25-liter container of water for a distance of 2-3 km for three or four times a day. To a few who are fortunate they can use wheelbarrows to carry 40-80 liters of water, usually twice a day. Carrying water by head or wheelbarrow is inducing some health problems, these containers are very heavy to most of us, I used to push 80 liters in a wheelbarrow now I am pushing 40 liters after I start developing backpains.* [Photo voice]

One of the study participants narrated that:

> *All household chores require water, such as cooking, cleaning and laundry; therefore, we women and girls are tasked to fetch water for our families.*

Fig. 6.4 Women fetching water in Chegutu Urban. (Source: Authors)

Fig. 6.5 Girls fetching water early in the morning in Chegutu town. (Source: Authors)

Likewise, most of the women respondents stated that water-carrying and doing the laundry were their most tiring jobs. Water shortages had an enormous bearing on women and girls as most had their normal daily routines altered to accommodate water fetching, which sometimes had to stretch from early morning until midnight. Figure 6.5 shows girls fetching water early in the morning in Chegutu town.

Fig. 6.6 Long queue at a borehole in Chegutu. (Source: Fieldwork)

> *These girls spend most of their time fetching water, there is nothing productive they are doing except spending time at the borehole, they come to the borehole three times a day; early in the morning around 4 am, then around 5 am and later in the afternoon after school. They rarely sleep. They do not even have time for homework or to rest. If water woes continue their future is slowly dying. They are being trained to be water fetchers.* [Photo story by Mrs Makombe]

In some instances, women went to boreholes with very long queues (Fig. 6.6), forcing some to wake up as early as two o'clock to queue at the local boreholes. Other women had to walk for long distances of about three kilometers to the nearby boreholes to fetch water. The school-going girls would spend several hours queuing for water as the queues would be very long. As a result, the education of girls has been highly affected. Many of them sometimes miss school or arrive late as they will be fetching water at home. Their studying time at school is also affected as they are forced to come home after school and proceed to the boreholes to fetch water. One representative of a local NGO narrated that:

> *Young women are affected to the extent of attending classes late and experiencing inadequate resting time, thereby leading to their poor performance at school as they will engage in the academic curriculum, already drained and tired. Cases of underage girls carrying heavy containers have become the norm, which is a form of child abuse as some households will be practising the so-called 'division of labor.*

To address some of the challenges related to borehole drilling, experts and policymakers advocate for stricter regulations on borehole drilling. Mr Mushore highlighted the importance of sustainable practices, noting that "unregulated drilling not only depletes groundwater reserves but also jeopardizes the quality of this essential resource." The government has been encouraged to implement measures such as licensing requirements for drilling companies, mandatory environmental impact assessments, and regular monitoring of aquifer levels. Communities have also begun adopting measures to safeguard groundwater resources. For instance, some neighbourhoods have established cooperative boreholes, where a single borehole serves multiple households, reducing the need for individual installations and minimizing over-extraction. Additionally, there is a growing emphasis on integrating borehole use with rainwater harvesting and graywater recycling to reduce the pressure on groundwater resources.

Despite some challenges, borehole drilling remains a vital strategy for enhancing water security in urban Zimbabwe. As Mrs Ncube, a small business owner in Gweru, explained:

> *Our borehole has been a lifeline. With it, we have been able to keep our operations running even when municipal water is unavailable.*

By combining technological innovation with sustainable practices and regulatory oversight, borehole drilling can continue to serve as a cornerstone of urban water management in Zimbabwe.

6.8.2 Use of Wells

Wells have long been a cornerstone of water provision, particularly in rural and peri-urban areas of Zimbabwe. With the growing challenges of water scarcity and unreliable municipal supplies, the use of wells has gained renewed significance as an affordable and practical water management strategy. Wells tap into shallow groundwater sources, providing communities with a readily accessible and cost-effective water supply. This method has become especially prevalent in areas where advanced solutions like boreholes are either financially or technically unfeasible.

The benefits of wells are numerous. Unlike boreholes, wells are relatively easy and inexpensive to construct, making them an accessible option for low-income households. In many cases, community members pool resources to dig and maintain shared wells, fostering a sense of cooperation and mutual support. Mr Dlamini, a farmer in Chegutu, shares his experience:

> *Our community well has been a lifeline for us. It provides water not just for drinking but also for irrigating our small gardens, especially during the dry season.*

Figure 6.7 shows a well servicing a home in Gweru town.

However, the effectiveness and sustainability of wells depend heavily on proper construction and maintenance. Unprotected wells are prone to contamination from surface runoff, particularly in areas with inadequate sanitation infrastructure. This contamination increases the risk of waterborne diseases such as cholera and dysentery, especially during the rainy season. Figure 6.8 illustrates a protected well in Chegutu Ward 12, showcasing the importance of safeguarding wells with proper covers and lining to prevent contamination.

To address these risks, organizations and local governments have promoted the construction of protected wells. These wells are designed with reinforced linings, covers, and sometimes hand pumps to minimize contamination and improve water safety. Despite these advancements, challenges remain. Wells are dependent on the availability of shallow groundwater, which can fluctuate due to seasonal variations

Fig. 6.7 Well in Gweru town. (Source: Authors)

Fig. 6.8 Protected well in Chegutu ward 12. (Source: Authors)

and over-extraction in nearby areas. In drought-prone regions, wells may run dry, leaving communities without reliable water access.

Regulations and best practices are essential to ensure the sustainable use of wells. Experts recommend regular monitoring of water quality and levels to prevent over-extraction and contamination. Mrs Banda, a community leader in Chegutu town, emphasized the importance of education and awareness:

> *We've seen how important it is to teach people about well maintenance. Simple actions like keeping the area around the well clean and using a bucket with a rope can make a big difference in preventing contamination.*

Integration with other water management strategies can also enhance the effectiveness of wells. For instance, communities are increasingly combining well use with harvesting rainwater to diversify their water sources and reduce dependence on a single supply. Greywater recycling systems can also complement well use by reducing overall water demand.

Despite its limitations, the use of wells remains a vital water management strategy in Zimbabwe. With proper safeguards and community cooperation, wells can continue to provide essential water access to underserved populations. As Ms Mutasa, a resident of a peri-urban area near Harare, explained:

> *For us, our well is more than just a water source—it's our lifeline. With a little support and knowledge, we can ensure it serves us for many years to come.*

By addressing challenges through education, technology, and regulation, wells can play a pivotal role in achieving water security for Zimbabwe's most vulnerable communities.

6.8.3 Reducing Water Use

One prominent approach identified in the urban communities to adapt to climate-induced water shortages is the conscious reduction of water consumption. This strategy involves the adoption of water conservation measures aimed at curbing excessive water intake and minimizing wastage. Through a combination of practical techniques and behavioral adjustments, households strive to optimize water usage and alleviate the strain on dwindling water resources.

One of the practical measures for water conservation is substituting conventional appliances with high-efficiency alternatives. Households have recognized the importance of upgrading their water-related appliances to more efficient models that consume less water without compromising functionality. For instance, replacing old, water-intensive washing machines with newer, high-efficiency models, significantly reduces water consumption per load. However, such adjustments are only possible with high-income earners. Mr Chikore, a homeowner in Bulawayo, shares his experience with substituting appliances:

> *When we switched to a new washing machine with better water efficiency, I was surprised by how much water we were able to save each month. It's a small change that adds up in the long run.*

Similarly, installing low-flow faucets and showerheads helps to regulate water flow, thereby conserving considerable amounts of water over time. Another proactive measure undertaken by households is the prompt detection and repair of water leaks within their premises. Leaky pipes, faucets, and toilets can collectively waste substantial volumes of water if left unattended. Mrs Ncube, a resident in Harare, emphasized the importance of repairing leaks promptly:

> *As soon as I notice even a small drip from the faucet, I make sure to get it fixed right away. Every drop counts, especially now with water shortages becoming more frequent.*

By vigilantly monitoring their water infrastructure and swiftly addressing any leaks or drips, households prevent unnecessary water loss and contribute to overall conservation efforts. Beyond technological interventions, households have embraced behavioral changes aimed at fostering a culture of frugality in water consumption. This involves instilling mindful habits such as turning off taps when not in immediate use, taking shorter showers, and refraining from unnecessary water usage during daily activities. Through education and awareness campaigns, communities are encouraged to adopt these simple yet impactful practices in their daily routines, thereby collectively reducing water demand and preserving precious water resources. Ms Moyo, a community organizer, highlights the significance of behavioral changes:

> *Educating people about simple habits like turning off the tap while brushing teeth or using a bucket to collect shower water for plants can make a big difference. It's about changing mindsets and habits for a more sustainable future.*

The reduction of water usage through the adoption of conservation measures and behavioral adjustments emerges as a proactive response to climate change-induced water shortages in urban Zimbabwe. By implementing these strategies at the household level, communities contribute to the broader effort of ensuring water security and resilience in the face of environmental challenges.

6.8.4 Utilizing Storage Tanks

In response to the erratic and often insufficient mains water supply exacerbated by climate change, families in urban Zimbabwe have also turned to using storage tanks as a proactive measure to mitigate water shortages. This strategy involves investing in rainwater harvesting systems and overhead tanks equipped with pumps, enabling households to collect and store precipitation for various purposes. Storage tanks are commonly used in both residential and commercial sectors in Zimbabwe to effectively manage water supply issues. These tanks allow users to store water during periods of availability, which can then be used during outages or restrictions. The installation of these tanks in cities like Bulawayo has been encouraged by municipal authorities, often supported by subsidies or incentives to reduce the upfront costs for households. By harnessing rainwater, families not only supplement their routine utility inflow during periods of scarcity but also ensure a sustainable water supply for non-potable applications such as gardening and laundry. Families have increasingly recognized the value of rainwater as a valuable resource that can alleviate dependence on unreliable mains water supply.

Rainwater harvesting systems, comprising gutters, downspouts, and storage tanks, are installed to capture and store rainwater runoff from rooftops. These systems are designed to efficiently channel rainwater into overhead tanks equipped with pumps, allowing for easy access and distribution when needed. Mr Nyathi, a homeowner in Mutare, shares his experience with rainwater harvesting systems:

> *Installing a rainwater harvesting system has been a game-changer for us. During the dry season when the mains supply is unreliable, we can rely on the water we've collected to keep our household running smoothly.*

However, storage tanks are usually beyond the reach of poor households in urban Zimbabwe. Figure 6.9 shows storage water tanks in Gweru town.

During periods of water scarcity or disruptions in mains water supply, stored rainwater serves as a reliable alternative source to meet essential household needs. By tapping into their reservoirs, families can maintain essential water usage for drinking, cooking, and sanitation, thus ensuring continuity in daily activities despite external water challenges. Mrs Sibanda, a resident in Gweru, highlighted the versatility of rainwater for non-potable applications:

> *We use the rainwater we collect for watering our vegetable garden and doing laundry. It's reassuring to know that we're not wasting precious mains water on tasks that don't require it.*

In addition to serving as a backup supply for potable water needs, collected rainwater is also utilized for non-potable applications within households. This includes watering gardens, washing clothes, and cleaning, where the quality of water is less critical. By diverting rainwater for such purposes, families effectively reduce the

Fig. 6.9 Storage water tanks in Gweru town. (Source: Authors)

strain on municipal water resources while embracing sustainable practices in water management. Mr Chuma, a community leader in Harare, emphasized the importance of storage tanks for water security:

> *Storage tanks equipped with pumps have become essential for urban households facing water shortages. They provide a buffer against irregular mains supply and empower families to take control of their water needs.*

The utilization of storage tanks, including rainwater harvesting systems and overhead tanks, emerges as a practical response to climate change-induced water shortages in urban Zimbabwe. By investing in these systems, families enhance their resilience to water scarcity while embracing sustainable water management practices that contribute to long-term environmental sustainability.

6.8.5 Adopting Climate Smart Agriculture

Some farmers in urban areas like Harare in Zimbabwe are embracing innovative cultivation techniques as a proactive measure to optimize agricultural productivity, in the face of escalating water shortages induced by climate change. Recognizing the imperative to adapt to prevailing moisture limitations, farmers are adopting climate-smart agricultural practices tailored to enhance resilience and sustainability in food production. These practices encompass a range of strategies, including agroforestry integration, mulching, drip irrigation, and the selection of drought-tolerant crop varieties, aimed at mitigating the impact of water scarcity while maximizing agricultural yields. Some farmers are incorporating trees and shrubs into their agricultural landscapes through agroforestry practices. Through intercropping trees with food crops, farmers benefit from increased soil moisture retention, reduced evaporation, and improved microclimate regulation. This integration not only conserves water but also enhances soil fertility and biodiversity, contributing to sustainable and resilient agricultural systems. Mr Moyo, a smallholder farmer in Harare, reflected on the benefits of agroforestry integration:

> *Integrating trees into our fields has made a noticeable difference in soil moisture levels. Even during dry spells, our crops have better access to water, and we've seen improvements in yields.*

The widespread adoption of mulching techniques further aids in moisture conservation within agricultural fields. Farmers apply organic materials such as crop residues, straw, or compost to the soil surface, forming a protective layer that reduces water evaporation, suppresses weed growth, and maintains soil moisture levels. Mulching not only conserves water but also promotes soil health and nutrient retention, thereby enhancing crop resilience to water stress. Innovative irrigation methods, such as drip irrigation, are increasingly being utilized by some farmers to deliver water directly to the root zone of crops with minimal wastage. Drip systems efficiently utilize water resources by applying precise amounts of water where it is needed most, thereby maximizing irrigation efficiency and minimizing water loss through evaporation or runoff. This technology enables farmers to sustainably manage water resources while optimizing crop yields in water-limited environments. Mrs Chikomo, a farmer in Bulawayo, shares her experience with drip irrigation:

> *Drip irrigation has transformed how we manage water on our farm. We're using less water overall, and our crops are healthier and more productive as a result.*

Farmers are proactively selecting and cultivating crop varieties that exhibit resilience to drought and water stress. Through drought-tolerant crops adapted to local climatic conditions, farmers reduce their vulnerability to water shortages and minimize yield losses during periods of limited rainfall. These resilient crop varieties are characterized by their ability to thrive under moisture-deficient conditions, thereby safeguarding food security and livelihoods in the face of climate variability. Mr Dube, an agricultural extension officer, emphasized the importance of drought-tolerant crop varieties:

> *By planting drought-tolerant varieties, farmers are building resilience against the impact of climate change. These crops can withstand water stress, ensuring food security even in challenging conditions.*

In summary, the adoption of climate-smart agricultural practices represents a proactive response to climate change-induced water shortages in urban Zimbabwe. Through these innovative techniques, farmers enhance agricultural resilience, conserve water resources, and sustainably manage food production systems in the face of evolving climatic conditions.

6.8.6 Water Reuse

Freshwater reserves in urban Zimbabwe are increasingly strained by climate change-induced water shortages, and the practice of water reuse has emerged as a sustainable solution to alleviate pressure on finite water resources. Water reuse involves treating wastewater to a safe quality and then reusing it for various purposes such as agriculture, industrial processes, or landscaping. While this approach is still being pioneered in Zimbabwe, some pilot projects have demonstrated its potential. For example, in Harare, treated wastewater has been used for irrigating public parks, although this practice is not yet widespread due to public health concerns and the need for more robust treatment systems (Thebe & Mangore, 2012). Dr Masunda, a water resource management expert, emphasized the importance of water reuse:

> *Water reuse is a critical strategy in sustainable water management, particularly in regions facing water scarcity. By repurposing graywater for auxiliary uses, we can significantly reduce the strain on freshwater reserves.*

Through the post-treatment of graywater, households and communities can repurpose wastewater for auxiliary uses, effectively reducing the demand for potable water and promoting conservation efforts.

Graywater, originating from sources such as sinks, showers, and laundry, undergoes rigorous posttreatment processes to remove contaminants and impurities. These treatment methods, which may include filtration, disinfection, and chemical treatment, render the graywater safe and suitable for reuse in non-potable applications. Through transforming graywater into recycled effluents, households can harness a valuable alternative water source for various tasks. Mr Dube, a community leader, highlighted the impact of water reuse on local communities:

> *In our community, water reuse has become a norm. Through using recycled water for tasks like toilet flushing and car washing, we're doing our part to conserve water and protect our environment.*

Recycled effluents find application in a variety of auxiliary tasks where the stringent quality standards of potable water are not necessary. One common application is toilet flushing, where recycled water can replace freshwater without compromising sanitation standards. Additionally, recycled effluents are utilized for car washing, providing an eco-friendly alternative to traditional water sources. Moreover, recycled water supports garden nurturing tasks, providing essential irrigation for plants while conserving precious freshwater reserves. Ms Chisango, an environmental scientist, discussed the benefits of recycled effluents for gardening:

> *Recycled effluents are rich in nutrients and can provide excellent irrigation for gardens. Through utilizing recycled water for gardening tasks, households can conserve freshwater resources while promoting healthy plant growth.*

Therefore, water reuse through the posttreatment of graywater offers a practical solution to alleviate pressure on freshwater reserves in urban Zimbabwe. By repurposing wastewater for auxiliary uses such as toilet flushing, car washing, and gardening, households and communities can conserve valuable freshwater resources, promote sustainability, and build resilience in the face of water scarcity.

6.8.7 Rainwater Harvesting

Communities in urban Zimbabwe are also increasingly turning to rainwater harvesting as a sustainable solution to augment their water supply in response to the growing challenges of water scarcity exacerbated by climate change. Rainwater harvesting is one of the most promising adaptation strategies, involving the collection and storage of rainwater for reuse. This approach has been particularly effective in residential areas where installation costs are relatively low, and the benefits include reduced dependence on municipal water supplies and alleviation of pressure on groundwater resources. The Government of Zimbabwe has supported these initiatives through educational campaigns and by providing technical assistance to households that wish to install rainwater harvesting systems (Munamati et al., 2016). Beyond simple tank installations, innovative approaches to rainwater harvesting are being implemented, including the delineation of rooftop catchment surfaces and the construction of subterranean reservoirs. These strategies aim to

Fig. 6.10 Rainwater Harvesting in Chegutu town. (Source: Authors)

maximize the capture efficiency and storage capacity of rainwater, thereby enhancing water security and resilience in the face of prolonged aridity phases.

Communities recognize the importance of maximizing the capture efficiency of rainwater to optimize water availability during dry periods. To achieve this, rooftops are strategically delineated as catchment surfaces to channel rainwater runoff into collection systems. By leveraging the expansive surface area of rooftops, communities can capture significant volumes of rainwater, minimizing runoff and maximizing water utilization. Mrs Moyo, a resident in Harare, highlighted the benefits of maximizing capture efficiency:

> *By optimizing our rooftop catchment surfaces, we've been able to capture more rainwater and reduce our reliance on municipal water sources. It's a simple yet effective way to enhance our water security.*

Figure 6.10 demonstrates rainwater harvesting in Chegutu town.

> *During rainy seasons we depend on rainwater harvesting. This is very easy, and time-saving. We only need widely open containers to harvest water such as dishes and buckets. This technique is seeming to be very helpful to a number of residents. We were advised to use harvested water for non-consumption purposes such as laundry and cleaning the houses. Asbestos roofs are deemed as unhealthy, so we are very cautious on it. However, we must have big tanks to store harvested water for use in future [Photo voice by Mrs Chibako]*

In addition to conventional above-ground tanks, communities are implementing innovative storage solutions such as subterranean reservoirs to store excess rainwater. These reservoirs, located underground, offer several advantages, including protection from evaporation and reduced visual impact on the landscape. By storing rainwater underground, communities can secure a contingency buffer against extended aridity phases, ensuring a reliable water supply even during periods of water scarcity. Mr Chikwava, a community leader in Bulawayo, discussed the importance of innovative storage solutions:

> *Subterranean reservoirs have been a game-changer for us. Not only do they provide ample storage capacity, but they also protect our water supply from evaporation, ensuring we have water available when we need it most.*

Rainwater harvesting represents a vital strategy for enhancing water security and resilience in urban Zimbabwe. By maximizing capture efficiency through rooftop catchment surfaces and implementing innovative storage solutions such as subterranean reservoirs, communities can effectively augment their water supply and mitigate the impacts of climate change-induced water shortages. As these practices continue to evolve and expand, they play a crucial role in building sustainable and resilient water management systems for the future.

6.8.8 Water Rationing and Scheduled Supply

Water rationing and scheduled supply have also become critical strategies for managing urban water shortages in Zimbabwe's major cities, including Harare and Bulawayo. As water demand continues to outstrip supply due to population growth, aging infrastructure, and the effects of climate change, these measures are employed by municipalities to systematically distribute limited water resources. While effective in prolonging water availability, water rationing presents numerous challenges that influence daily life, hygiene, and sanitation, particularly in vulnerable communities.

The rationale behind water rationing is to manage scarce resources by allocating water in intervals, often on a rotational schedule. In Harare, for instance, certain residential areas may receive water for a few hours or days each week, depending on the availability of municipal supplies. This approach helps prevent the complete depletion of reservoirs during drought periods or when infrastructure failures occur. Mr Chikafa, a resident of Harare, explained the mixed feelings associated with rationing:

> *While it's frustrating to have water only a few times a week, at least we know when to expect it and can plan accordingly.*

However, water rationing is not without its drawbacks. Prolonged cuts often lead to disruptions in daily routines, forcing households to rely on alternative sources such as boreholes or water vendors. This can significantly increase household expenses and exacerbate inequalities, as wealthier families are better able to afford supplementary solutions than low-income residents. The lack of reliable water can also compromise hygiene and sanitation, contributing to outbreaks of waterborne diseases such as cholera and typhoid. Mrs Dube, a mother of four in Bulawayo, lamented:

> *When the taps run dry, it becomes impossible to maintain proper hygiene, especially for large families.*

Municipalities often justify water rationing as a temporary measure to stabilize supply while addressing systemic issues. Yet, critics argued that the practice highlights deeper problems, including insufficient investment in water infrastructure and poor governance. Mrs Chikodza emphasized this point by stating,

> *Water rationing, though necessary in the short term, is symptomatic of broader failures to modernize and expand water systems in line with urban growth.*

Efforts to address these systemic issues have included initiatives to rehabilitate dams and pipelines, but progress has been slow due to financial and logistical constraints.

To mitigate the negative effects of rationing, some communities have adopted adaptive strategies. For instance, rainwater harvesting has gained popularity as a supplementary water source during dry periods. Additionally, local authorities have encouraged water conservation practices, such as fixing leaks and reducing unnecessary usage. In Bulawayo, community leader Mr Ndlovu, advocated for collective responsibility:

> *We all have a role to play in using water wisely. Even small actions, like reusing graywater, can make a difference when the supply is limited.*

Despite its challenges, water rationing remains a vital tool for managing urban water scarcity in Zimbabwe. As municipalities continue to grapple with the complexities of water supply and demand, the success of this strategy depends on effective communication, equitable implementation, and long-term investments in infrastructure. By combining these efforts, cities can better navigate the immediate pressures of water scarcity while laying the groundwork for a more sustainable future.

6.9 Evaluation of Urban Water Adaptation Strategies

The effectiveness of urban water adaptation strategies in the face of climate change varies widely. Borehole drilling, for instance, while providing a reliable alternative to municipal water, poses long-term sustainability issues due to the potential for over-extraction. Rainwater harvesting, on the other hand, offers a more sustainable supplement to household water supplies, though its effectiveness is limited by seasonal variability and the initial cost of installation. Sustainability concerns particularly regarding groundwater use and water rationing, highlight the need for IWRM approaches. Such approaches would consider all water sources and their interactions, aiming to sustainably balance water availability with ecological and human needs. The scalability of the strategies is also a critical consideration. While individual households can relatively easily implement rainwater harvesting and water-saving technologies, scaling these solutions to serve larger populations requires substantial investment and coordination. This is particularly challenging under Zimbabwe's current economic conditions, where funding for large-scale infrastructure projects is limited.

Table 6.3 provides an overview of various water management strategies, assessing their effectiveness, sustainability, and the challenges they face, which can be crucial for policy-makers and planners in urban water governance.

While various local adaptation strategies provide temporary relief from water shortages in Zimbabwean cities, their long-term success requires integration into a comprehensive, sustainable water management framework that addresses both supply and demand aspects of water governance. This approach should be supported by robust policy initiatives, community engagement, and international cooperation to ensure water security in the face of ongoing climate challenges.

Table 6.3 Assessment of the effectiveness and sustainability of urban water management strategies

Strategy	Effectiveness	Sustainability	Challenges
Water-use reduction	Moderate	Moderate	Requires behavioral changes and public awareness programs
Rainwater harvesting	Moderate	Moderate	Dependent on rainfall; requires initial setup and maintenance
Storage tanks	Moderate	Moderate	Requires initial investment; feasibility varies by household income and space availability
Climate-smart agriculture	High	Moderate	Needs ongoing education and support; initial learning curve for adaptation
Water reuse	Moderate	Low to moderate	Requires significant infrastructure investment; needs to address public health concerns
Borehole drilling	Moderate to high	Moderate	Risk of groundwater depletion; unregulated drilling may lead to contamination and aquifer degradation
Use of wells	Moderate	Low to moderate	Susceptible to contamination; dependent on shallow groundwater availability; may run dry during droughts
Water rationing and scheduled supply	Low to moderate	Low	Disruptive to daily life; affects hygiene and sanitation; highlights infrastructure and governance issues

6.10 Recommendations for Enhancing Urban Water Resilience in the Age of Climate Change

Urban areas in Zimbabwe, as in many parts of the world, are increasingly challenged by the multifaceted impacts of climate change on water resources. To enhance resilience and ensure sustainable water supply and quality, comprehensive strategies are required. The following recommendations detail approaches to fortify urban water systems against the backdrop of climate change (Table 6.4).

6.11 Conclusion

This chapter explored urban water adaptation in the context of climate change in Zimbabwe. The chapter revealed that climate change caused complex and multidimensional challenges for secure water service delivery in urban Zimbabwe. The chapter highlighted that persistent water scarcity, deteriorating water quality, and the increasing frequency of extreme weather events are exacerbating existing issues within urban water management in urban Zimbabwe. As Zimbabwe continues to urbanize and grapple with these impacts, the importance of transforming water governance to enhance resilience and sustainability has never been more apparent. One of the key findings of this analysis is the paramount need to shift from traditional water management strategies to more integrated, adaptive, and resilient approaches. It is essential that urban water governance not only incorporates strategies for adaptation to immediate climate stresses but also builds long-term resilience against future uncertainties. This involves enhancing the capacity of water infrastructure to withstand climate extremes and ensuring that water policies are flexible and responsive to changing environmental conditions and scientific insights.

Moreover, the concept of hydrocitizenship has emerged as crucial in the age of climate change. It emphasizes the need for active involvement of citizens in water governance, promoting a sense of shared responsibility and collective action towards water sustainability. In the urban areas of Zimbabwe, fostering hydrocitizenship can empower communities, particularly those in vulnerable and underserved neighborhoods, to participate meaningfully in shaping water policies and practices that directly affect their lives and livelihoods. The chapter also calls upon policymakers, urban planners, and community leaders to take decisive action. There is a pressing need to review and revise existing water management frameworks to ensure they are robust enough to cope with the complexities introduced by climate change. Investments should be prioritized in upgrading ageing infrastructure, implementing green infrastructure solutions, and promoting water conservation technologies. Moreover, policies should facilitate the adoption of innovative water management approaches such as rainwater harvesting, water reuse, and the sustainable extraction of groundwater.

Table 6.4 Strategic recommendations for enhancing urban water resilience in the age of climate change

Category	Recommendations	Details
Integrated water resource management (IWRM)	Develop policies integrating water management with urban planning, promote water-efficient technologies, enhance institutional capacity	Implement policies that integrate surface water and groundwater management with urban development, encourage water-efficient appliances in residential and industrial sectors, and enhance the capacity of water governance institutions
Green infrastructure	Implement permeable pavements, green roofs, rain gardens, conserve urban wetlands and riparian zones, promote urban forestry	Use natural systems to manage stormwater, enhance groundwater recharge, reduce urban heat island effects, and improve city aesthetics and biodiversity
Diversification of water supply	Develop alternative water sources like rainwater harvesting, reclaimed wastewater, desalination	Reduce dependence on single water sources by developing technologies and systems that provide alternative water sources
Upgrading and retrofitting infrastructure	Replace aging infrastructure, reinforce critical water systems, install advanced metering	Upgrade and retrofit water supply and sanitation infrastructure to reduce leakages, improve efficiency, and withstand extreme weather events
Enhancing forecasting and monitoring systems	Implement advanced hydrological models, utilize geographic information systems (GIS) and remote sensing, and develop early warning systems	Use modern tools and technologies for better forecasting of water availability and risks, and to monitor the health of water resources and infrastructure in real-time
Adaptive management practices	Establish feedback mechanisms, promote research and development in water technologies, and encourage pilot projects	Adapt management strategies based on observed outcomes and evolving conditions, test new ideas and technologies, and continually assess effectiveness
Public awareness and participation	Conduct public education campaigns, involve communities in decision-making, develop school programs	Increase public knowledge and engagement in water conservation and climate change impacts, involve citizens in water management planning and decision-making
Policy and governance strengthening	Update water policies and regulations, strengthen institutional capacities, ensure transparency	Review and enforce water laws and regulations to remain relevant and effective under climate change scenarios, build institutional capacity, and ensure governance practices are transparent and accountable
Economic instruments	Implement tiered water pricing, offer subsidies for water-saving measures, establish water markets	Use economic tools to encourage water conservation, support investments in sustainable water management, and facilitate the efficient allocation of water resources
Collaboration and partnerships	Foster partnerships across sectors, participate in networks, encourage community-based management	Leverage resources, expertise, and innovation through collaborations between government, the private sector, academia, and communities; share knowledge and best practices internationally

Community engagement and education are essential in cultivating a culture of water conservation and management. Urban planners and community leaders are urged to work collaboratively to integrate public input into water governance processes, ensuring that decisions are inclusive and equitable. This collaborative approach can harness the collective expertise and resources necessary to address the multifaceted challenges posed by climate change. In conclusion, as Zimbabwe faces the growing impacts of climate change, the resilience of its urban water systems is critical to the nation's overall sustainability and prosperity. By prioritizing adaptation and resilience, enhancing hydrocitizenship, and fostering robust, collaborative governance, Zimbabwe can ensure that its urban water systems are not only sustainable but also resilient enough to effectively serve future generations.

To sum up the recommendations, policymakers should do the following:

- Craft robust policies and ensure effective implementation: Policymakers must enact and enforce legislation to ensure that climate change adaptation measures are embedded in urban water governance.
- Build strong institutions: Strong institutions capable of delivering water services and responding to the impacts of climate change are critical for enhancing urban water sector resilience and adaptation to climate change.
- Foster community engagement and prioritize stakeholder collaboration: Encourage active community participation in decision-making processes and foster stakeholder collaboration to build shared responsibility and commitment to climate change adaptation efforts.
- Invest in research and innovation: Continuous investment in research and innovation is required to generate new ideas and solutions to tackle emerging challenges brought about by climate change.
- Increase funding and allocate resources wisely: Increased funding and wise resource allocation are essential to support climate change adaptation efforts in the urban water sector.

By embracing these recommendations, Zimbabwe can bolster its ability to adapt to climate change in the urban water sector, thereby safeguarding the lives and livelihoods of millions of urban citizens.

References

Adedeji, O., Reuben, O., & Olatoye, O. (2014). Global climate change. *Journal of Geoscience and Environment Protection, 2*(2), 114–122. https://doi.org/10.4236/gep.2014.22016

Allan, R. P., Arias, P. A., Berger, S., Canadell, J. G., Cassou, C., Chen, D., et al. (2023). Intergovernmental Panel on Climate Change (IPCC). Summary for policymakers. In V. P. Masson-Delmotte, P. Zhai, & A. Pirani (Eds.), *Climate change 2021: The physical science basis. Contribution of Working Group I to the Sixth Assessment Report of the Intergovernmental Panel on Climate Change* (pp. 3–32). Cambridge University Press. https://doi.org/10.1017/9781009157896.001

Bhanye, J. I., Mangara, F., Matamanda, A. R., & Kachena, L. (2023). *COVID-19 lockdowns and the urban poor in Harare, Zimbabwe*. Springer. https://doi.org/10.1007/978-3-031-41669-9

Chikozho, C. (2010). Applied social research and action priorities for adaptation to climate change and rainfall variability in the rainfed agricultural sector of Zimbabwe. *Physics and Chemistry of the Earth, Parts A/B/C, 35*(13–14), 780–790. https://doi.org/10.1016/j.pce.2010.07.006

Cutter, S. L., Emrich, C. T., Gall, M., & Reeves, R. (2018). Flash flood risk and the paradox of urban development. *Natural Hazards Review, 19*(1), Article 05017005. https://doi.org/10.1061/(ASCE)NH.1527-6996.0000268

Diffenbaugh, N. S., & Burke, M. (2019). Global warming has increased global economic inequality. *Proceedings of the National Academy of Sciences, 116*(20), 9808–9813. https://doi.org/10.1073/pnas.1816020116

Folke, C., Carpenter, S. R., Walker, B., Scheffer, M., Chapin, T., & Rockström, J. (2010). Resilience thinking: Integrating resilience, adaptability and transformability. *Ecology and Society, 15*(4) https://www.jstor.org/stable/26268226

Hallegatte, S., Green, C., Nicholls, R. J., & Corfee-Morlot, J. (2013). Future flood losses in major coastal cities. *Nature Climate Change, 3*(9), 802–806. https://doi.org/10.1038/nclimate1979

Hayes, S., Desha, C., Burke, M., Gibb, M., & Chester, M. (2019). Leveraging socio-ecological resilience theory to build climate resilience in transport infrastructure. *Transport Reviews, 39*(5), 677–699. https://doi.org/10.1080/01441647.2019.1612480

IPCC (Intergovernmental Panel on Climate Change). (2021). *Climate change 2021: The physical science basis, contribution of Working Group I to the Sixth Assessment Report of the Intergovernmental Panel on Climate Change*. Cambridge University Press. https://doi.org/10.1017/9781009157896

Johnson, M. T., & Munshi-South, J. (2017). Evolution of life in urban environments. *Science, 358*(6363), Article eaam8327. https://doi.org/10.1126/science.aam8327

Jones, V., Whitehouse, S., McEwen, L., Williams, S., & Gorell, B. L. (2021). Promoting water efficiency and hydrocitizenship in young people's learning about drought risk in a temperate maritime country. *Water, 13*(18), Article 2599. https://doi.org/10.3390/w13182599

Kaika, M. (2004). *City of flows: modernity, nature, and the city*. Routledge.

Kløve, B., Ala-Aho, P., Bertrand, G., Gurdak, J. J., Kupfersberger, H., Kværner, J., et al. (2014). Climate change impacts on groundwater and dependent ecosystems. *Journal of Hydrology, 518*, 250–266. https://doi.org/10.1016/j.jhydrol.2013.06.037

Manzungu, E., & Mabiza, C. (2004). Status of water governance in urban areas in Zimbabwe: Some preliminary observations from Bulawayo and Harare. *Physics and Chemistry of the Earth, 29*(15–18), 1167–1172. https://doi.org/10.1016/j.pce.2004.09.003

Marmorek, D., Nelitz, M., Eyzaguirre, J., Murray, C., & Alexander, C. (2019). Adaptive management and climate change adaptation: Two mutually beneficial areas of practice. *Journal of the American Water Resources Association, 55*(4), 881–905. https://essa.com/wp-content/uploads/2019/07/Marmorek_et_al._in-press.-AM-and-CCA-two-mutually-beneficial-areas-of-practice-.pdf

Masson-Delmotte, V. P., Zhai, P., Pirani, S. L., Connors, C., Péan, S., Berger, N., et al. (2021). Summary for policymakers. In *Climate change 2021: The physical science basis. The Working Group I contribution to the sixth assessment report of the Intergovernmental Panel on Climate Change*.

Mgwele, A., Bhanye, J. I., & Mocwagae, K. S. (2024). Water and sanitation for urban health: A gender perspective on impacts and coping strategies in Mangaung Metropolitan Municipality. In A. R. Matamanda & V. Nel (Eds.), *Sustainable development goals and urban health: strides, challenges and way forward for poor neighborhoods* (pp. 49–67). Springer.

Mohajerani, A., Bakaric, J., & Jeffrey-Bailey, T. (2017). The urban heat island effect, its causes, and mitigation, with reference to the thermal properties of asphalt concrete. *Journal of Environmental Management, 197*, 522–538. https://doi.org/10.1016/j.jenvman.2017.03.095

Mukuhlani, T., & Nyamupingidza, M. T. (2014). Water scarcity in communities, coping strategies and mitigation measures: The case of Bulawayo. *Journal of Sustainable Development, 7*(1). https://doi.org/10.5539/jsd.v7n1p144

Munamati, M., Nhapi, I., & Misi, S. (2016). Exploring the determinants of sanitation success in Sub-Saharan Africa. *Water Research, 103*, 435–443. https://doi.org/10.1016/j.watres.2016.07.030

Ndebele-Murisa, M. R., & Mubaya, C. P. (2019). Decision making and climate resilience in the water sector of Harare. *Policy Brief.* https://www.fractal.org.za/wp-content/uploads/2019/02/Harare-Policy-Brief.pdf

Ndunguru, M. G., & Hoko, Z. (2016). Assessment of water loss in Harare, Zimbabwe. *Journal of Water, Sanitation and Hygiene for Development, 6*(4), 519–533. https://doi.org/10.2166/washdev.2016.157

Pahl-Wostl, C., Craps, M., Dewulf, A., Mostert, E., Tabara, D., & Taillieu, T. (2007). Social learning and water resources management. *Ecology and Society, 12*(2) https://www.jstor.org/stable/26267868

Perry, S. J., McGregor, S., Sen Gupta, A., England, M. H., & Maher, N. (2020). Projected late 21st century changes to the regional impacts of the El Niño-Southern oscillation. *Climate Dynamics, 54*, 395–412. https://doi.org/10.1007/s00382-019-05006-6

Putsoane, T., Bhanye, J. I., & Matamanda, A. (2024). Extreme weather events and health inequalities: Exploring vulnerability and resilience in marginalized communities. *Developments in Environmental Science, 15*, 225–248. https://doi.org/10.1016/B978-0-443-21948-1.00011-X

Rammile, S., Bhanye, J. I., Lehobo, T., & Hugo, M. (2024). Building resilient urban infrastructure and innovation for urban public health: A case study of Bloemfontein, South Africa. In A. R. Matamanda & V. Nel (Eds.), *Sustainable development goals and urban health: Strides, challenges and way forward for poor neighborhoods* (pp. 69–96). Springer.

Ren, Z., Fu, Y., Dong, Y., Zhang, P., & He, X. (2022). Rapid urbanization and climate change significantly contribute to worsening urban human thermal comfort: A national 183-city, 26-year study in China. *Urban Climate, 43*, Article 101154. https://doi.org/10.1016/j.uclim.2022.101154

Reniko, G., Phonchi-Tshekiso, N. D., Mogomotsi, P. K., Mogomotsi, G. E., & Chirefu, H. (2020). Analysis of factors influencing domestic water consumption in Karoi town, Zimbabwe. *Journal of Sustainable Development, 13*(2), 82–97. https://doi.org/10.5539/jsd.v13n2p82

Revi, A., Satterthwaite, D., Aragón-Durand, F., Corfee-Morlot, J., Kiunsi, R. B., Pelling, M., et al. (2014). Towards transformative adaptation in cities: The IPCC's fifth assessment. *Environment and Urbanization, 26*(1), 11–28. https://doi.org/10.1177/0956247814523539

Scott-Bottoms, S., & Roe, M. (2020). Who is a hydrocitizen? The use of dialogic arts methods as a research tool with water professionals in West Yorkshire, UK. *Local Environment, 25*(4), 273–289. https://doi.org/10.1080/13549839.2020.1732897

Shayamunda, R. H. (2024). Access to basic water and sanitation infrastructure among women in African informal settlements: Implications for health. *Developments in Environmental Science, 15*, 53–76. https://doi.org/10.1016/B978-0-443-21948-1.00004-2

Takagi, H., Esteban, M., Mikami, T., & Fujii, D. (2016). Projection of coastal floods in 2050 Jakarta. *Urban Climate, 17*, 135–145. https://doi.org/10.1016/j.uclim.2016.05.003

Thebe, T. A., & Mangore, E. N. (2012) Wastewater production, treatment and use in Zimbabwe. Department of Civil and Water Engineering, National University of Science and Technology, Bulawayo.

Toriro, P. (2024). People-centred initiatives as a response to infrastructure collapse in Harare. In *Urban infrastructure in Zimbabwe: Departures, divergences and convergences* (pp. 181–199). Cham.

United Nations. (2018). *2018 revision of world urbanization prospects*. United Nations, Department of Economic and Social Affairs.

Weeks, J. R. (2010). Defining urban areas. In T. Rashed & C. Jürgens (Eds.), *Remote sensing of urban and suburban areas* (pp. 33–45). Springer.

WMO (World Meteorological Organization). (2023). *State of the global climate 2023*. World Meteorological Organization. WMO No. 1347.

Zhao, Q., Yu, P., Mahendran, R., Huang, W., Gao, Y., Yang, Z., et al. (2022). Global climate change and human health: Pathways and possible solutions. *Eco-Environment and Health, 1*(2), 53–62. https://doi.org/10.1016/j.eehl.2022.04.004

Zinyama, T. (2021). Local government, decentralisation, devolution, and service delivery in Zimbabwe. In M. Khosrow-Pour (Ed.), *Encyclopedia of information science and technology* (5th ed., pp. 1597–1611). IGI Global. https://doi.org/10.4018/978-1-7998-3479-3.ch109

Chapter 7
Nexus Between Gender and Urban Water Supply: A Feminist Political Perspective

Abstract In most African cities, women and young girls dominate water management at the household level. Yet, their dominance wanes at the higher levels of water management. Using the feminist political perspective, this chapter explores strategies to address inequality and women's social, cultural, and political subordination in Zimbabwe's urban water supply sector. The chapter argues that concerted efforts between men and women in addressing water woes in urban Zimbabwe are mutually beneficial. However, the chapter revealed that women's voices remain weak in terms of urban water supply management. Most women and young girls are subjected to distressing experiences gathering water for the household. Suburbs in various urban centers show some discrepancies regarding the burden of collecting water for the household. In low-income suburbs, women have a more significant role in managing household water issues. However, in the middle-income suburbs, women's involvement in gathering water for the household is less strenuous, especially in families with automobiles and supportive male counterparts. The chapter recommends a national quota system stipulating that at least half of the members in any water committee or water department within municipalities should be females. This strategy can potentially increase women's participation in decisions that affect their daily lives—water supply. Furthermore, behavior change programs are necessary to encourage males to increase their involvement in water provision and management at the household level.

Keywords Gender roles · Water governance · Urban water supply · Water woes · Participation

7.1 Introduction

A long list of factors are linked to climate change in African cities, including institutional weaknesses, financial shortages and increasing scarcity. Access to clean drinking water is a challenge in most communities. Globally, approximately 600 million people lack access to improved drinking water sources, most of which are

A. R. Matamanda et al., *Urban Water Governance in Postcolonial Zimbabwe*, Local and Urban Governance, https://doi.org/10.1007/978-3-031-96973-7_7

people living in poverty in the rural areas of developing countries (United Nations, 2015). Lack of access to clean drinking water has a tremendous impact on the workloads of women and men in countries in the Global South (Mgwele et al., 2024; Shayamunda, 2024). Women are often in charge of collecting water, cleaning, cooking, and caring for the sick (Gambe, 2019). These chores drastically limit the time spent by women on paid work and leisure and, in the case of girls, the time for educational pursuits. Experiences vary across regions. In sub-Saharan Africa, women are usually charged with water collection in 65% of rural households and men in 10%. Women collect water in 33% of urban households compared to 10% of men (United Nations, 2015). The time spent by women collecting water is highest in sub-Saharan Africa, with a round trip averaging 33 and 25 min in rural and urban areas compared to 21 and 19 min in Asia's rural and urban areas (United Nations, 2015).

The nexus between gender and urban water supply is well documented, especially in the Global South (Mgwele et al., 2024; Shayamunda, 2024). Typical experiences in some parts of Asia, Latin America and Africa show that women and young girls are charged with the role of water provision at the household level (Abrahams et al., 2011; Gambe, 2015, 2019; Khosla & Pearl, 2003). This emanates from women's cultural, social, economic, and political subordination. In most African cultures and societies, women and girls collect water for household uses such as cooking, cleaning, health and hygiene (Khosla & Pearl, 2003). For example, in South Africa, the nexus between gender and water poverty mainly results from the unequal gender division of labor, with women responsible for water collection (Abrahams et al., 2011).

This unequal division of labor adversely affects women's participation in income-generating activities, perpetuating their economic subordination. Their predicament is worsened by their limited participation and involvement in politics. In some African countries, active political participation by women is often associated with loose morals, which limit their ambitions to take up political positions. However, this is gradually changing as both women and men become enlightened and empowered to embrace change, leaving behind archaic practices that disadvantaged women.

In Zimbabwe, water poverty is regarded mainly as a challenge for women since they are the water managers at the household level. Notwithstanding this, men are increasingly involved in assisting women in collecting water for the household (Gambe, 2015, 2019). The burden experienced by women varies according to the place of residence. Water scarcity is more pronounced in the rural than urban areas. The average distance travelled and time spent collecting water is more in the rural than urban areas. Women in rural areas endure the challenges of water collection with little or no help from men. In comparison, urban areas present a slightly different scenario. Women have different experiences depending on the suburbs they live in. Urban women with the most significant burden are those in unplanned/partially planned settlements not connected to a reticulated water supply system. Their experiences are comparable to those in low-income residential areas with intermittent piped water supply. Therefore, using the feminist political perspective, this chapter explores strategies to address inequality and women's social, cultural, and political subordination in the urban water supply sector. The study had two specific objectives: to assess the sociocultural, economic and political subordination of women and young girls in Zimbabwe's urban water supply sector and to explore policy strategies that can be adopted to address women's subordination in the urban water supply sector.

7.2 Feminist Political Perspective

The feminist political perspective is inherently diverse (Sunstein, 1989). However, most feminist writings address issues related to the social subordination of women and the possibility of reformulating political theory and practice by eliminating gender biases (Sunstein, 1989). In fact, as a political movement, feminism seeks to address inequality and all forms of subordination of women (social, cultural, economic and political) (Tucker, 2015). The essence of feminist politics is to end the domination of women through transforming the institutions and practices that support women's subordination. Tucker (2015:6) noted that the feminist political theory posits that "different social and economic locations produce different views of the world. Moreover, the more privileged one's economic and social position, the less likely one will have a true understanding of reality." This chapter adopts this perspective in assessing how economic and social standing influences one's understanding of water scarcity and its implications on gender. Furthermore, the chapter discusses how institutions and practices in Zimbabwe can be reformed to end women's subordination in the urban water supply sector.

7.2.1 Socio-cultural Subordination

Zimbabwe is predominantly a patriarchal society with a gender division of roles and responsibilities, especially at the household level. However, this has changed over time, and households exhibit different experiences. While water management at the household level is mainly ascribed to women and young girls, who are expected to fetch water, cook, clean and take care of the health and hygiene in the family, men and young boys are slowly getting involved. The conventional wisdom that water collection interferes with the efforts of women and young girls to engage in paid work and attend school, is increasingly challenged by emerging trends, at least in urban areas. Perceptions that emerged from the interviews include:

> *My husband is unemployed, and I am the one who goes to work. Despite this, my husband does not do household chores, especially fetching water. He expects me to do that because he believes women should do it. I do it for my husband's respect but feel it is burdensome.* (Respondent 1)

> *I am a full-time housewife, but my husband does provide most of the water we need for household use. He uses our single-cab pickup truck to bring us water from his workplace. He fetches the water from the company's borehole.* (Respondent 2)

> *I treat my children the same regardless of gender. I have two sons and two daughters, and they all take turns to fetch water for the households. The boys are more prominent in household water supply because the girls are sometimes occupied with other household chores.* (Respondent 3)

The above sentiments show that the family unit as an institution can either perpetuate or discontinue the subordination of women in terms of household water supply and management. However, attitude and beliefs matter. The three responses show how different attitudes and beliefs can produce varied outcomes. In households still deeply involved in cultural practices, women and young girls are burdened with a myriad of household chores even when men and young boys can assist. Yet, some households have moved away from abusive cultural practices and have embraced gender equality. This is exemplified by the sentiments of Respondents 2 and 3.

It emerged that the gender division of labor is sometimes influenced by cultural marriage beliefs and practices. Dowry (lobola) payment subjects women to various types of subordination. Culturally, a married man considers himself to have acquired a helper (wife) who, besides bearing children, is supposed to do the household chores for him. Women are, therefore, ascribed responsibilities that include cooking, cleaning the house, doing the laundry and other associated household chores. On the other hand, a married man is expected to work and care for the family, mainly in terms of providing monetary resources needed by the household. The household chores ascribed to women are heavy water consumers, which is how they become household water managers. As an institution, marriage can be a source of women's subordination. People's understanding, perceptions and beliefs about marriage can reinforce or discontinue the subordination of women.

Religion also plays a role in women's social and cultural subordination in Zimbabwe. The study found that some church practices perpetuate women's domination. Key issues emanating from the study include:

> *In our church, women are taught to submit to their husbands. I am used to it because I grew up in this church. I feel some duties should be done by women and others by men, so I do not have a problem being responsible for fetching water for my family.* (Respondent 4)

> *Our church allows men to have many wives. My husband is the head of the family, and as wives, we are responsible for all the household chores, including water collection. It is considered disrespectful for women to ask their husbands to assist with water collection.* (Respondent 5)

Doctrines taught by different churches influence and prepare women to accept subordination. For example, some apostolic churches support polygamy and teach women to be submissive. Their teachings influence how the division of labor is done in the families, with women and young girls mainly assigned to do all household chores, including water collection. Other churches do not allow their members to participate in elections, which limits women's ability to be voted into political offices or select politicians who can champion their needs. The church as an institution can be a medium for positive change. Churches should not preach and perpetuate the subordination of women but play a role in their emancipation.

Social subordination of women in urban areas partly emanates from the failure of local municipalities to meet the residents' demand for various types of infrastructure and services. In cases where piped water supply is unavailable, women and young girls take up the responsibility of water collection. Water rationing has become the norm in Zimbabwe, with suburbs going for weeks, months, or years without receiving piped water supplies. Residents have responded differently to this, depending on their economic status. The study revealed that private boreholes connected to houses' water systems are common in high-income and some middle-income suburbs. In comparison, hand-dug wells, communal boreholes, dams and rivers or streams are commonly used in low-income and unplanned or partially planned suburbs.

This distribution of alternative water sources indicated in the preceding paragraph shows that women in high-income and parts of middle-income suburbs are not as burdened by water scarcity as those in low-income and unplanned settlements. Most urban poor have limited access to water (Wilk & Jonsson, 2013). In unplanned or partially planned settlements, women are raped during the night while on their trips to fetch water for their families (Matamanda et al., 2024). These rape cases go unreported because the victims fear their marriages will break if their husbands know of their ordeal. Young girls are physically and sexually harassed while queuing to fetch water from communal boreholes. Male borehole marshals are the main perpetrators.

7.2.2 Political Subordination

For a long time, men have dominated politics in Zimbabwe at the local, provincial and national levels. However, this has been changing as laws and policies of the country have undergone revisions to promote gender equality in political structures and institutions. This section briefly explains women's subordination in Zimbabwe, exploring ways to address this, especially in the urban water supply sector.

Historically, some rules and policies implemented in Zimbabwe had a demeaning effect on women. For example, some popular police operations implemented in the country targeting women include *Operation Chinyavada* (Scorpion, 1983), *Operation Chipo Chiroorwa* (Ladies Get Married, 2007), *Operation Chengetedza Hunhu* (Maintain your Dignity, 2011) and *Operation Dyira Bonus Kumba* (Spend your Annual Bonus at Home, 2012) (Benyera & Nyere, 2015). These operations were purported to clear the streets of prostitutes. Women found walking the streets of urban centers after 18:00 without the company of men were arrested and charged for contravening Section 8 of the Criminal Law (Codification and Reform Law) Act, Chapter 9:23, which forbids loitering to solicit sex. Surprisingly, men were not arrested during these operations. Thus, the operations exhibited patriarchal tendencies that reduced the independence of women.

7.2.3 Women's Participation in Politics

The experiences explained in the previous paragraph also affect women's participation in political activities. Although more than 50% of the 6.5 million registered voters for the 2023 election were women, only 1 of the 11 presidential candidates was a woman. Besides that, only 11% of the candidates for the national assembly posts were women, down from 14% in the 2018 election (Matiashe, 2023). Likewise, the percentage of women in the national assembly and senate has declined since the 2013 election, as shown in Table 7.1.

Although the COZ sets aside 60 of the 270 seats in parliament for women through proportional representation (Matiashe, 2023), this is not enough to push the total percentage of women in parliament to a 50–50 representation stipulated in the COZ. The political parties, as institutions, are not doing much to push up the number of women who participate in the general elections. In as much as the election candidates are selected through primary elections, political parties can adopt a deliberate policy that encourages women to hold political office. Without political power, it remains difficult for women to influence policy and challenge their political subordination.

The study found that political domination of women is still rife in local authority elections in Zimbabwe. This political structure is essential for women because it influences local municipalities' service delivery (including water supply). Yet, most of the elected officials, especially in urban local authorities, are not pushing hard enough for the interests of women. Enhancing this political structure is not easy because it involves changing the attitude of voters, who need to elect officials who can effectively represent their interests in council meetings regardless of gender. It emerged that in most cases, women who can effectively represent the interests of their peers are not elected to office because politics is not considered a profession for women. One of the candidates of the 2023 local authority election in Chitungwiza (the third largest city in Zimbabwe) indicated that:

> *I believe the voting process was not about selecting a person who can represent the people's interests. It was all about which party the candidate represented and, to an extent, gender. I could have done better if the electorate was considering the ability to deliver. However, the voters in my ward still think that politics is for men and women running for political office are weak and of loose morals, but this is a myth.* (Respondent 8)

Table 7.1 Percentage of women in Zimbabwe's national assembly and senate

Year	Percentage of women politicians	
	National assembly	Senate
2013	33	48
2018	31	44
2023	30.7	–

Adapted from Madsen and Mtero (2023)

Few women run for political office due to namė-calling and negative criticism, particularly from male rivalries. Two factors compound the situation. First, elections in Zimbabwe are mostly characterized by violence, discouraging women with political ambition, especially those without strong backing from their male counterparts. As such, the political platform is tilted in favor of men, perpetuating women's subordination. Second, the economic resources required for the nomination of candidates and to run a successful campaign are out of reach for many aspiring women politicians, especially those in opposition parties struggling to acquire campaign funds.

Without a strong representation of women politicians in local municipalities, improvements in the access to piped water supply will remain a dream. Sometimes, male politicians (who dominate local municipalities) do not seem to understand or appreciate the burden posed on women and young girls by water shortages, or they simply do not care. As noted by one of the women in Chitungwiza after the 2023 elections:

> *... izvozvi hatina mvura. Vanhu veZANU PF vakavhara zvibhorani zvizhinji zviri muno muZengeza nokuti vakadyiwa pama elections. Kwanzi zvibhorani ndezve ZANU saka imi chererwai zvibhorani ne CCC yamakavhotera.*
>
> *[...we are currently not able to access the water from most communal boreholes in Zengeza because they were locked up by angry and frustrated ZANU–PF supporters who could not handle their loss in the 2023 harmonised elections. They claim that the boreholes were provided by ZANU PF, so the people of Zengeza should ask the party they voted for CCC [Citizens Coalition for Change] to drill boreholes for them]* – Respondent 9

The behavior of the ZANU–PF supporters in Chitungwiza shows intolerance and a lack of concern for the welfare of the people, especially women. Water is a right enshrined in the COZ, and no citizen should deny another access to the precious resource.

The claim that the boreholes were provided by ZANU–PF is without merit. Some boreholes were donated to Chitungwiza under the Presidential Borehole Drilling Scheme. Others were drilled by former Member of Parliament using constituency development funds. Because of this, the boreholes belong to the residents of Zengeza (Chitungwiza) not ZANU–PF supporters alone. The Presidential Borehole Drilling Scheme is a government, not a political party program, so ZANU–PF supporters had no right to lock up the boreholes. Even if the ZANU–PF party solely donated the boreholes to the people of Zengeza, the supporters cannot deny others access to water because their candidate(s) lost in an election. Such behavior is counter-productive and shows how service provision is politicized in local municipalities. Service provision is considered a favor offered to residents in exchange for votes. Yet, services should be provided without political connotations. The post-2023 election experiences in Zengeza have motivated women to participate in politics as a means to challenge their subordination. They recognized that the city's persistent water crises are unlikely to be resolved as long as the political landscape remains dominated by men, some of whom resort to punitive actions like locking up water points after losing an election.

7.2.4 Women's Participation in Local Municipalities' Water Departments

The participation of women and young girls in the higher levels of the water supply and management sector in Zimbabwe is still limited. This explains why water poverty in the country is a burden associated with women and young girls (Gambe, 2019). The water management departments in most local authorities are mainly dominated by men, with a few women involved, holding lower and less influential positions. One of the reasons is the limited number of women and girls enrolling for engineering courses at universities and polytechnics. Because of this, most engineers heading water supply departments in the majority of Zimbabwe's local municipalities are men. The domination of men in these departments affects the water supply and management decisions made by local municipalities. Turning around this situation entails encouraging women and young girls to participate in science, technology, engineering and mathematics subjects. This can increase the number of female engineers responsible for managing water authorities in the country. Apart from that, a quota system is required in all water departments and committees in the country, stipulating the percentage of women in such structures and the positions they should potentially hold. A 50% quota will likely empower women and strengthen their voice in urban water supply systems.

7.2.5 Economic Subordination

Economic hardships in Zimbabwe have increased the number of unemployed people and forced most urban residents into informal sector employment. For those in informal employment, competition for survival is stiff, especially in the vegetable and fruit vending business. Women in this business endure unequal opportunities with men due to the strain posed by household water collection activities. When water supplies are reconnected in the early morning (for example, at 02:00), mostly women and young girls wake up to fetch water for storage and do laundry before supplies are disconnected (Gambe, 2019). These chores negatively affect women's vending activities because they fail to get quality produce at the market for reselling. In comparison, their male counterparts in the vending business can get to the market early and get the best produce. Consequently, water collection activities negatively affect women's ability to compete economically with men.

The prolonged blackouts experienced in the country have also affected women in informal businesses. The blackouts have caused water disconnections spanning over two weeks in some suburbs. Women have to make plans to get water for domestic use. Cities like Bulawayo and Masvingo are typical examples. In Bulawayo, prolonged power outages cause regular water supply interruptions as reservoir levels decrease (Mpofu, 2022). Severe load shedding has also worsened water shortages in Masvingo (Mukaro, 2023). Owing to the power outages, women in these cities spend most of their time fetching water for the household. Yet, the time can be used to further the small businesses that give them income. With improved income, their economic dependence on their husbands is reduced.

The study revealed that power outages in Zimbabwe also affect women's vending businesses. The respondents reported varied effects:

> *I am involved in a peanut butter-making business. The unplanned and prolonged power blackouts have affected me negatively. Last night, power was reconnected around midnight. I woke up, switched on my machine, and started processing peanut butter. However, within 20 to 25 minutes, power was disconnected. I was distressed because I did not produce the number of bottles I expected, and some of my produce was spoilt due to the unplanned shutdown of the machine.* (Respondent 6)

> *I am a producer and bulk supplier of freezits. My business is currently going down due to electricity problems in the country. I recently lost two deep freezers due to a bad electric current. I cannot fix them now because I do not have enough funds.* (Respondent 7)

It emerged that earnings from these businesses are usually used to purchase household groceries and water for domestic use. The women pay push-cart operators who offer water supply services. The push-cart operators can fetch as many as 20 or 25 20-liter buckets and deliver the water at their homes (Gambe, 2019). This significantly reduces the energy and time spent by women collecting water for the household. The women can rechannel the saved energy and time into economic activities that make them economically independent. Thus, with their small businesses thriving, women can finance water collection activities and provide water for the household without being overly dependent on their husbands for money. However, institutions like the Zimbabwe Electricity Supply Authority make this difficult. Power outages may be inevitable, but the institution should produce load-shedding schedules that enable residents to plan accordingly, including those in business.

7.3 Policy Recommendations

Significant reforms are needed in the institutions and practices perpetuating women's sociocultural, political and economic subordination. Although reforms are difficult to effect in institutions or deeply ingrained traditions, this chapter suggests that social, political, economic, and administrative changes are required. However, the basic family unit presents a base to propagate all the anticipated changes.

7.3.1 Reforms in Social Institutions

The family unit presents a base to formulate and implement social, political and economic changes. Change of attitude and perceptions should start at the household level before being exported to other social groupings. The management of water collection at the household level should promote emancipation rather than perpetuate women's subordination. Positive behavior and respect for equality can be taught at the household level, and then members spread the change by influencing others in their networks.

Churches are influential institutions in Zimbabwe. As such, churches can play an essential role in reducing women's sociocultural subordination. For example, churches can work on revising their doctrines to preach women's emancipation. The starting point entails adjusting oppressive doctrines and dogmas. However, this may be easier for Indigenous churches than orthodox ones. Notwithstanding this, those in dire need of change are the former compared to the latter. The founders of Indigenous churches have much influence over their followers, which is a fertile ground for initiating change. Government and NGOs representing women's interests can arrange 'positive change' meetings or workshops with founders of Indigenous churches seeking to affect a paradigm shift in how the churches perceive and treat women. The essence is to influence the churches to uphold and respect the rights of women and young girls. Achieving this provides a platform for women to enjoy equal access to social, economic and political resources. More importantly, the strategy can reduce women's and young girls' burden of household water collection. It should become the whole family's responsibility, regardless of gender, to fetch and manage water for the household.

7.3.2 Reforming Political Institutions

The nexus between water provision and urban politics in Zimbabwe is intrinsic. Urban politics decide who gets what and when. As water managers at the household level, women must be actively involved in urban politics if they expect meaningful improvements in the water supply sector. Although the political terrain is not conducive for women to freely participate, women should push for reforms to level the playing field. First, female politicians should change public opinion and prove that women can be good political leaders. The electorate must witness and be convinced that leadership is not defined by gender. Selection of leaders should be based on the candidate's ability to address the electorate's needs. Despite being dominated by male politicians, poor governance is widespread in most local municipalities. This suggests that by holding other factors constant, gender does not influence the performance of political leaders.

Second, political parties should consider developing a long-term plan to field an equal number of male and female candidates. The fact that only 11% of the candidates sought political office in the 2023 national assembly election suggests that the political parties in the country are still patriarchal. As such, these institutions should start encouraging and promoting female party members to take political office. One way of doing so is to stipulate that any primary election conducted by the parties to elect its representative for any election should have both male and female candidates. In doing so, the parties should avoid imposing candidates even if they are women because this is against the democratic principles of electing political leadership.

Third, the security forces in the country need to tighten security before, during and after elections. Curbing violence in the political arena will likely encourage women to brave up and seek political offices. Women should not relax and expect to get political posts through the quota system. They must demand political space and fight for their needs and rights, including improved water service delivery. Male politicians should play their part in opening up political spaces for women to participate and fight for their cause. Finally, women's economic empowerment is necessary to reduce or end their political subordination. Economically emancipated women can better participate in urban politics than those financially dependent on their husbands or male counterparts. Thus, women's economic freedom is associated with their political emancipation.

7.3.3 Improved Access to Economic Opportunities

The burden of fetching water for the household is heavy on women without or with limited access to economic opportunities. As indicated previously, water woes have a different impact on women in different economic groups. In Zimbabwe, the burden is heavy on women staying in low-income suburbs and unplanned settlements compared to those in middle- and high-income suburbs. There have been efforts by the Ministry of Women Affairs, Community, Small and Medium Enterprises Development to empower micro, small and medium enterprises, some of which are owned by women. However, these efforts are not enough. This Ministry should increase the support given to women in low-income and unplanned settlements, especially those with the potential to run businesses. The support could be in business and financial management training, or the provision of capital and machinery required in their business adventures.

Government, private firms and NGOs can create associations of women entrepreneurs that track the progress of those who have benefitted through training or machinery provided by any of the three institutions. The associations can also encourage other women to engage in businesses to liberate themselves from economic subordination. In terms of urban water supply, this chapter argues that economically empowered women can effectively lessen the burden associated with acute water shortages. Two factors support this argument. First, economically active women can pay for drilling boreholes or wells within their plots—a common practice in middle- and some low-income settlements. With a well at home, the time and energy women and girls spend on fetching water for the household is significantly reduced. Second, women who cannot afford to drill boreholes or wells within their compounds can pay pushcart operators to service their water supply needs.

7.3.4 Reforms in Water Management Institutions

The study found that women have significant participation in water point committees and social groupings formed at the community level to address water issues. Yet, this is not the same in the water departments of local municipalities. The number of women in these water departments is still low (Gambe & Dube, 2015). Some action to turn around the status quo has been put in motion countrywide, namely encouraging the enrolment of women and young girls in science, technology, engineering and mathematics subjects. This is expected to improve the number of women in water or civil engineering disciplines and participation in central and local government water departments. The approach will not only reduce the economic but also the social and political subordination of women. An alternative approach would encourage on-the-job training and upskilling for women in the water departments.

7.4 Conclusion

Breaking women's political, economic, social and cultural subordination is a complex task that does not happen overnight. It has been embedded in urban and rural communities for so long and was propagated from the traditional culture which perpetuated patriarchy. The chapter used the feminist political perspective to explore strategies to address inequality and women's social, cultural, and political subordination in Zimbabwe's urban water supply sector. It addressed two objectives: to assess the sociocultural, economic and political subordination of women and young girls in Zimbabwe's urban water supply sector; and to explore policy strategies that can be adopted to address women's subordination in the urban water supply sector. Significant progress has been made to break away from unfair cultural practices. The belief that gathering water for the household is a duty of women and young girls is slowly disappearing. Cases of men and young boys collecting water for the family are evident in most urban communities. This shows a paradigm shift in the right direction, even though practices differ from one household to the other.

Ending women's domination in the urban water supply sector can be achieved by transforming the institutions and practices that support women's subordination. As previously discussed, these institutions include the basic family unit, churches, political parties and local municipalities. Enhancing women's voice and participation in these institutions can be achieved by improving their income and education. In the high-income suburbs, most women are not bothered about the current water woes in the country. They either have personal boreholes or purchase bulk water delivered into private reservoirs within their plots. Similar cases are also evident in some middle-income suburbs. Those who cannot afford to drill boreholes resort to protected wells that supply the bulk of their needed water. Water for drinking is either bought or fetched from nearby boreholes.

Women in low-income suburbs and unplanned settlements are mainly affected by water scarcity and have limited options to overcome the burden imposed upon them by acute water shortages. For example, taps in some of Zimbabwe's low-income suburbs have been dry for more than 3 years. Without money to drill either boreholes or protected wells, women in these settlements depend on communal boreholes, unprotected wells and nearby streams for their daily water supply. Gathering water from these sources takes a lot of time and energy. Moreover, it exposes women and young girls to various forms of abuse. However, the number of boreholes, protected wells, and water tanks is slowly increasing in some low-income areas, though local municipalities do not approve them. The plot/stand sizes of these settlements are so small, ranging from 150 m^2 to 200 m^2 in most cases. As such, there is no provision for drilling boreholes or digging wells within stands of this size.

Women with high education levels have better chances of getting employed and earning an income, which gives them some form of freedom from subordination. The study found that women and young girls from wealthy families with higher education levels are afforded the same opportunities as men and young boys within the household. Therefore, providing water for the household is every member's responsibility, regardless of gender. In contrast, women and young girls in poor and uneducated families have most of their time consumed by water collection duties. Furthermore, they lack time to study or engage in paid work. Thus, sociocultural, economic and political subordination of women in the water sector is widespread in low-income and unplanned settlements where the most vulnerable people reside.

References

Abrahams, Y., Mhlongo, S., & Napo, V. (2011). A gendered analysis of water and sanitation services policies and programmes in South Africa: 2006–2010. *Agenda: Empowering Women for Gender Equity, 25*(2), 71–79. https://www.jstor.org/stable/41321417

Benyera, E., & Nyere, C. (2015). An exploration of the impact of Zimbabwe's 2005 operation Murambatsvina on women and children. *Gender & Behaviour, 13*(1), 6522–6534. https://uir.unisa.ac.za/handle/10500/18774

Gambe, T. R. (2015). Prospects of prepaid smart water metering in Harare, Zimbabwe. *African Journal of Science, Technology, Innovation and Development, 7*(4), 236–246. https://doi.org/10.1080/20421338.2015.1081762

Gambe, T. R. (2019). The gender dimensions of water poverty: Exploring water shortages in Chitungwiza. *Journal of Poverty, 23*(2), 105–122. https://doi.org/10.1080/10875549.2018.1517399

Gambe, T. R., & Dube, K. (2015). Water woes in Harare, Zimbabwe: Rethinking the implications on gender and policy. *International Journal of Innovative Research and Development, 4*(6), 390–397. https://www.academia.edu/19295940/Water_Woes_in_Harare_Zimbabwe_Rethinking_the_Implications_on_Gender_and_Policy

Khosla, P., & Pearl, R. (2003). *Gender, water and poverty: Key issues, governments commitments and actions for sustainable development.* http://www.unwater.org/downloads/untapped_eng.pdf

Madsen, D. H., & Mtero, S. (2023). Zimbabwe's election was a fight between men—women are sidelined in politics despite quotas. *The Conversation*. 30 August. https://theconversation.

com/zimbabwes-election-was-a-fight-between-men-womenare-sidelined-in-politics-despite-quotas-212131. Accessed 16 March 2025.

Matamanda, A. R., Mphambukeli, T. N., & Chirisa, I. (2024). Exploring water-gender-health nexus in human settlements in Hopley, Harare. *Cities & Health, 8*(1), 44–58.

Matiashe, F. (2023). Another Zimbabwe election cycle reveals decline of women in politics. *Al Jazeera*. 22 August. https://www.aljazeera.com/features/2023/8/22/another-zimbabwe-election-cycle-reveals-decline-of-women-in-politics. Accessed 15 March 2025.

Mgwele, A., Bhanye, J. I., & Mocwagae, K. S. (2024). Water and sanitation for urban health: A gender perspective on impacts and coping strategies in Mangaung Metropolitan Municipality. In A. R. Matamanda & V. Nel (Eds.), *Sustainable development goals and urban health: Strides, challenges and way forward for poor neighborhoods* (pp. 49–67). Springer.

Mpofu, M. (2022). All Bulawayo suburbs to go without water indefinitely. *Chronicle*. 30 December. https://www.chronicle.co.zw/all-bulawayo-suburbs-to-go-without-water-indefinitely/. Accessed 16 March 2025.

Mukaro, B. (2023). Water woes giving us headaches—Masvingo town Clerk. *Masvingo Mirror*. 4 April. https://masvingomirror.com/water-woes-giving-us-headaches-masvingo-town-clerk/. Accessed 19 March 2025.

Shayamunda, R. H. (2024). Access to basic water and sanitation infrastructure among women in African informal settlements: implications for health. *Developments in Environmental Science, 15*, 53–76. https://doi.org/10.1016/B978-0-443-21948-1.00004-2

Sunstein, C. R. (1989). Introduction: notes on feminist political thought. *Ethics, 99*(2), 219–228. https://www.jstor.org/stable/2381432

Tucker, E. (2015). Feminist political theory. In M. T. Gibbons (Ed.), *The encyclopedia of political thought* (pp. 1–13). Wiley-Blackwell.

United Nations. (2015). *The world's women 2015: Trends and statistics*. United Nations, Department of Economic and Social Affairs, Statistics Division. https://unstats.un.org/unsd/gender/downloads/WorldsWomen2015_report.pdf

Wilk, J., & Jonsson, A. C. (2013). From water poverty to water prosperity—A more participatory approach to studying local water resources management. *Water Resources Management, 27*(3), 695–713. https://doi.org/10.1007/s11269-012-0209-8

Chapter 8
Infrastructure Financing and Technological Advancements for Urban Water Supply

Abstract The financing of different types of infrastructure has become a core issue in Africa, considering its high urbanization rates. An acute shortage of water infrastructure in Zimbabwe is one of the primary causes of water scarcity in most cities and towns. This chapter explores the innovative financing mechanisms and technological advancements that can help ease urban water shortages. The study found that the limited participation of the private sector in urban water supply inhibits private financing in water infrastructure. Central and local government institutions dominate the water supply sector and operate on a cost-recovery basis. Some cities have embarked on a drive to reduce the water purification cost to create extra water supply funds. The strategy involves synergies between municipalities and local firms working towards replacing imported chemicals with locally produced water purification chemicals. Similar synergies are recommended, especially in finding sustainable financing mechanisms that can improve investments in water infrastructure. Meanwhile, urban households should consider using low-cost household purification techniques, including boiling water, aqua tablets and water filters that have become common in Africa's low-income communities. These household purification methods can guarantee the consumption of clean and safe water.

Keywords Water demand · Water finance · Privatization · Urbanization · Urban water supply

8.1 Introduction

Water infrastructure provision in African local municipalities has become an uphill task (Mgwele et al., 2024). The failure to provide the infrastructure mainly emanates from high urbanization rates and other factors like corruption and poor governance (Dodman et al., 2017; Rammile et al., 2024). In Zimbabwe, most cities are experiencing water woes caused by a series of intertwined factors (Gambe, 2015, 2019; Musemwa, 2010; Nel & Berry, 1992; Nhapi, 2009). In the capital, Harare, the daily water supply is less than 50% of the daily demand (Gambe, 2013). The city's

A. R. Matamanda et al., *Urban Water Governance in Postcolonial Zimbabwe*, Local and Urban Governance, https://doi.org/10.1007/978-3-031-96973-7_8

water source, Lake Chivero, is heavily polluted, requiring approximately eight chemicals to be purified to the level of human consumption. The water distribution network is worn out, and this has caused a high percentage of non-revenue water (Gambe, 2015). In Bulawayo, the second largest city, water scarcity emanates from the unavailability of nearby sustainable water sources (Nel & Berry, 1992). The Zambezi water project, meant to supply Bulawayo with bulk water, has been on the cards for a long time, but no significant progress has been made.

In the city of Chitungwiza, daily water demand outstrips supply. The city does not have enough bulk water resources to meet the needs of the residents. Therefore, the city purchases bulk water from Harare, which is also failing to satisfy its daily water demand (Gambe, 2019). The water distribution network in Chitungwiza has also been worn out and has contributed to water shortages. Unplanned settlements on top of water infrastructure have also worsened the problem (Matamanda et al., 2020). The city's failure to sustainably manage the sewage problem has contributed to increased pollution in natural water systems. The situation in the city of Masvingo is not different from other Zimbabwean cities. The water infrastructure shortage contributes to the municipality's failure to meet the daily water demand. Partially planned residential developments in and around the city are not connected to the reticulated water supply. Communal water points are still in use in some suburbs like Mucheke.

The common factor in the cases mentioned above is a failure to find sustainable sources of water infrastructure financing in Zimbabwe's local municipalities. The water sector is dominated by public institutions that mainly rely on central government grants and subsidies. In addition, local municipalities also raise revenue for water supply and management through rates and taxes. Yet, these sources of finance are not enough to cater for the water supply needs of urban areas This chapter explores the innovative financing mechanisms and technological advancements that can help ease urban water shortages. The specific research questions for this chapter are as follows:

- What are the factors behind the water woes in Zimbabwean municipalities?
- What technological advancements have been made in the urban water supply sector?
- What policy strategies can be implemented to improve urban water supply in Zimbabwe's urban municipalities?

The chapter proceeds with a review of factors behind water woes in Zimbabwean cities, and an evaluation of technological advancements in the water sector, followed by a discussion of policy recommendations and the conclusion.

8.2 Factors Behind Water Woes in Zimbabwe

Water scarcity in Zimbabwean cities emanates from a multiplicity of factors, including a limited number of bulk water sources, worn-out water purification plants and distribution systems, high pollution levels in water sources, political interference in the running of local authorities and the growth of unplanned settlements. All these challenges are associated with financing issues. However, cases vary across cities, as demonstrated in the subsequent subsections.

8.2.1 Shortages of Bulk Water Infrastructure

Several cities in Zimbabwe do not have bulk water sources to meet the residents' current and future demands. For example, water shortages in cities like Bulawayo and Chitungwiza are linked to a lack of nearby surface water bodies to meet residents' needs. Factors behind this abound, though financial constraints remain central.

Climatic conditions in the Matabeleland region are associated with low rainfall, affecting water availability in Bulawayo (Mukuhlani & Nyamupingidza, 2014; Musemwa, 2006). The water available in the city's dams is insufficient to meet the residents' demand throughout the year. The prominent proposed solution is the construction of a 450 km pipeline that draws water from the Zambezi River and channels it to reservoirs in Bulawayo (Musemwa, 2006; Nel & Berry, 1992). Because of the length of the required pipeline and the scope of work, this project has not yet seen the light of day. Some scholars have attributed this delay to national politicking (Mukuhlani & Nyamupingidza, 2014; Musemwa, 2006; Nel & Berry, 1992; Nhapi, 2009). Despite signing the unity accord, the Harare-based Zimbabwe African National Unity Patriotic Front (ZANU–PF) government has been accused of withholding funds towards the development of Bulawayo—the base of the Zimbabwe African People's Union. The central government has used this strategy to whip in-line cities controlled by opposition parties (Musemwa, 2006).

Located a few kilometers from Harare, Chitungwiza is relatively small in geographical coverage. The city developed as a satellite settlement in 1981 and has since been experiencing different types of infrastructural challenges. Chitungwiza does not have a reliable surface water body, so it depends on Harare for bulk water supply. Specifically, the former purchases approximately 28–30 ML water daily against its demand of around 45 ML (Gambe, 2019). Plans are underway to construct Muda Dam, but progress has been stalled by acute financial shortages. NGOs and the presidential borehole drilling program have provided relief to the residents by donating boreholes to the city. Of the 20 communal boreholes drilled under the presidential borehole drilling scheme, six have been completed and are functional (Madzimure, 2022).

8.2.2 *Worn-out Water Plants and Distribution Networks*

The study established that the water treatment plants, and distribution networks of most Zimbabwean cities are old and worn out. Most water treatment plants have outlived their design life, and their daily output cannot meet the residents' water needs. The same applies to the distribution networks, which are associated with high non-revenue water emanating from underground leakages. Against this backdrop, financial limitations have hamstrung the efforts by Zimbabwean municipalities to expand and maintain their water treatment plants and distribution networks. Because of this, water scarcity in cities has become a common phenomenon.

In cities like Harare and Mutare, financial doldrums had a negative impact on the effective operation and maintenance of the water supply systems (Bhanye et al., 2023). In the former, the main water treatment plant (Morton Jaffrey) needed expansion and refurbishment. Major maintenance works done in the recent past included refurbishing a part of the Morton Jaffrey treatment plant and replacing iron with plastic pipes in some sections of the city. However, these refurbishments have fallen short of the work required for the city to increase its daily water supply. The predicament is almost the same in other cities. In the latter, non-revenue water has remained high because of underground leaks and pipe bursts (Muleya, 2023). Therefore, non-revenue water has remained high in some cities and relatively lower in others (Table 8.1).

In general, the study revealed that non-revenue water in most cities results from burst pipes not attended for a long time, underground leakages, and water theft emanating from either vandalism of fire hydrants or bypassing of water meters (Gambe, 2013). The problem is compounded by non-functional water meters in most cities, though efforts to replace them are evident. Harare is slowly replacing conventional meters with smart water meters, starting with high-income areas with a sustained water supply (Gambe, 2015). In Chitungwiza, the study established that some houses do not have water meters, and it is therefore difficult to determine their water consumption patterns. In most cases, the water meters were stolen, and the households have not been able to replace them.

Table 8.1 Population and water supply in selected cities

City	Population 1975	Population 1990	Population 2000	Population 2015	Daily supply (mL)	Daily demand (mL)	Non-revenue water
Harare	765,398	1,322,061	1,573,937	1,934,205	700	1400	60%
Mutare	69,674	115,861	131,089	159,616	125	130	60%
Gweru	73,975	120,708	139,426	164,533	36	120	57%
Masvingo	40,385	67,022	76,077	90,601	30	45	41%
Bulawayo	501,159	688,479	676,741	665,971	155	165	35%
Chitungwiza	183,994	302,336	366,622	448,363	30	45	30%

Source: Authors creation using data from expert interviews, Bulawayo 24 News (2018); Global Human Settlements Urban Centre Database (2015), Gambe (2013, 2015); Moyo-Ndlovu (2023); Mubaiwa (2023); Mukaro (2023); Muleya (2023); World Bank (2013)

The study established that the percentage of non-revenue water is also pushed up by defective water billing systems in some cities. Water billing, based on estimates, can result in over- or under-charging. In the case of the former, the council loses potential revenue; in the latter, defaults increase as residents protest against unfair charges. The latter case affects the relationship between local municipalities and residents, resulting in conflicts that are difficult to resolve.

8.2.3 Pollution of Primary Water Sources

High pollution levels in the cities' primary surface water bodies have contributed to water scarcity in Zimbabwe. The sources of pollution include industrial wastes, chemicals used in urban agricultural activities, and raw sewage deposited in natural river systems. Challenges emanating from pollution vary from one city to the other. Harare is located within the drainage basin of its primary water source, and this explains why the capital's main water source (Lake Chivero) is heavily polluted. Industrial wastes, agrochemicals used in fields, oil leaks from vehicles, and raw sewage find their way into Lake Chivero through the Mukuvisi River that drains runoff from the city center and other surrounding suburbs. High pollution levels have significantly increased the city's water purification costs. Instead of three chemicals like other cities, Harare requires, on average, five to eight chemicals to treat the water to the recommended standard (Gambe, 2013). The water purification process is, therefore, more expensive in Harare than in other cities. Pollution of natural water bodies is not unique to Harare but a common occurrence in other cities. For example, Masvingo City allegedly disposed of raw sewage into two rivers that are tributaries of Lake Mutirikwi, the city's primary water source (Maponga, 2022). Besides killing aquatic life, such incidences make water purification costly.

Financial challenges in Zimbabwe's local municipalities have contributed to the failure to adequately treat sewage before being disposed of in the natural river system. This is the case of Masvingo. The city has been struggling to upgrade its sewage treatment plant due to financial challenges. As such, the current treatment plant cannot adequately purify sewage. Cases of dead fish in Lake Mutirikwi have been widespread, and the Environmental Management Agency fined the municipality for polluting the lake (Kusemachibi, 2022). A similar case is evident in Chitungwiza, a city known for struggling with sewage problems. The city is known as *kuma sewage*, meaning a place popularly known for sewage overflows (Gambe, 2019). Water pollution has been high because the city failed to address sewage problems.

8.2.4 Politics of Urban Water Supply

Literature has generally shown the nexus between water scarcity in Zimbabwean cities and political interferences in the governance of local municipalities. The country's urban water crisis is linked to the central government's interference in water provision decisions made by local municipalities, especially those controlled by opposition parties (Musemwa, 2006, 2010; Nel & Berry, 1992). Political alienation is considered as one of the causes of acute water shortages in Bulawayo (Musemwa, 2006; Nel & Berry, 1992). In the post-independence era, the Harare-based ZANU–PF government considered Bulawayo, the base of the opposition (the Zimbabwe African People's Union), as an errant city. Therefore, political differences between the two parties were associated with the alienation of Bulawayo, as the central government limits water development projects in the second-largest city (Musemwa, 2006; Nel & Berry, 1992). Although Bulawayo Metropolis is still under the control of the opposition party, mainly the CCC, the relationship between the city and central government seems to have improved compared to the period immediately after independence. The partnership of the central government and the African Development Bank has invested US$37 million in the rehabilitation of the city's water supply system through the Bulawayo Water and Sewerage Services Improvement Programme launched in 2016 (Moyo-Ndlovu, 2023). This project stabilized the water situation in Bulawayo.

The alienation of cities controlled by opposition parties is meant to achieve political submission. Like Bulawayo, Harare is under the control of opposition parties, mainly the CCC. Yet, political intervention has hamstrung water management in the city (Nhapi, 2009). The central government reverses some strategic decisions undertaken by Harare City Council through the Ministry of Local Government and Urban Housing, which has the mandate to oversee the operations of local authorities in Zimbabwe. Service delivery has become poor due to the discord in the city's governance.

8.2.5 City Expansion Patterns

Unplanned expansion in some Zimbabwean cities has affected the provision of municipal infrastructure. Informal settlements have emerged in and around cities, increasing the demand for water services. The emergence of such settlements has, in some cases, interfered with existing infrastructure. For example, in Chitungwiza the extent of informality includes construction on top of sewerage trunk lines and under electricity lines, construction on wetlands and areas surrounding cemeteries. This has increased sewerage overflows in the city which have increased the pollution of water sources, including wells used by the majority in the city. Peri-urban developments have of late been on the increase but without adequate planning on where the settlements will get water, sewerage and energy services. The demand for infrastructure associated with unplanned expansion of cities is high since most of the settlements are spatially fragmented (Matamanda et al., 2020). Local municipalities do not afford to provide the required infrastructure and this creates shortages.

8.3 Technological Advancements in Water Supply

The technological advancements that have occurred in the urban water supply sector in Zimbabwe have mainly been smart water metering and the use of GIS in the monitoring of water distribution networks.

8.3.1 Smart Water Metering in Zimbabwe

Some Zimbabwean municipalities have adopted smart water meters but have not seen widespread use. This limited adoption is attributed to financial constraints, lack of expertise, viability challenges, and low social acceptability. Proponents argued that smart meters can reduce non-revenue water and enhance water revenue collection efficiency. However, in most Zimbabwean cities, non-revenue water has significantly increased (see Table 8.1), primarily due to non-functional conventional water meters, underground leaks, water theft, and burst pipes.

8.3.1.1 Financial Constraints

Smart water metering requires a considerable capital outlay that depends on the functions needed from them. The meters can remotely record water consumption and automatically send data to the water utility. Furthermore, smart meters can detect leakages and help water utilities to quickly identify leakage points within the water distribution systems. Depending on their functions, smart water meters are expensive to install in neighbourhoods; therefore, most municipalities in Zimbabwe cannot afford the installation charges. Large cities like Harare have undertaken a pilot project with 2000 m installed in the central business district area (Gambe, 2015). A US$144 million bank loan was used to fund the project. It is not easy for small municipalities in Zimbabwe to get loans of this magnitude. Therefore, the affordability of this technology has remained out of reach for many local municipalities.

8.3.1.2 Lack of Expertise

The installation of smart water metering in Zimbabwean municipalities has also been dragged by the lack of expertise required to install and maintain the technology. Brain drain has been a significant challenge as highly qualified personnel relocate to other countries with better economic opportunities. Most municipalities are left without key personnel required to sustain smart water metering systems. Others cannot attract highly skilled personnel due to poor and below-regional-average remuneration.

8.3.1.3 Viability Concerns

Smart water metering benefits systems with a high water service level provision, that is, an uninterrupted 24/7 supply of water with quantities of 100 L per capita per day and above. Attaining such service levels is rare in Zimbabwean municipalities. Instead, an intermediate water service level is typical, which comprises a mixture of household and communal access points. For the intermediate water service level, water quantities are below 100 L per capita per day, and supplies are sometimes interrupted owing to water demand management practices.

In most cities, residents do not get tap water daily but can go for more than a month without supplies. Some parts of Msasa Park in Harare have gone without water for over a year (Gambe, 2013). The current power outages experienced in the country have worsened the situation, and cities like Bulawayo and Masvingo regularly experience water supply interruption owing to low water levels in the reservoirs (Mpofu, 2022; Mukaro, 2023). Installation of smart water meters in cities, towns, or suburbs experiencing acute water shortages is uneconomical since the installation costs far outweigh the revenue collected. Consequently, smart water metering remains a technology used in some suburbs and not others.

8.3.1.4 Social Concerns

The installation of smart water meters in Zimbabwe faced resistance from residents who associated the technology with the commercialization of water. The residents argued that water is a fundamental right guaranteed in the COZ and should not be commercialized (Gambe, 2015). Furthermore, the technology was against the social norms of sharing water in communities. As a precious resource, urban residents, especially in low-income suburbs, share water with those without supplies, free of charge. Smart water meters threaten this norm as households become cost-conscious. Sharing water with other households becomes a cost that needs to be settled. Thus, residents regard smart meters as a technology which promotes individualism. The other concern was water accessibility in low-income areas. The economic malaise in Zimbabwe has made life difficult for middle- and low-income households struggling to earn a living in urban centers. Therefore, the commercialization of water by installing smart meters can negatively affect the water accessibility of urban households that are already struggling to settle monthly water bills.

8.3.2 Low-Cost Water Purification Technologies

It emerged that acute water shortages in Zimbabwean urban centers have pushed residents to turn to groundwater. Boreholes connected to the household water supply systems are typical in high-income areas, while unprotected wells are widespread in low-income suburbs and unplanned/partially planned settlements.

Furthermore, communal boreholes have been drilled in water-stressed suburbs to give residents an alternative water source. However, the communal boreholes fall short of the water demand in these suburbs. This is why unprotected wells have become a common water source. Yet, water from these wells is not fit for human consumption. Apart from that, some communal boreholes are drying up with others being contaminated. For example, Harare decommissioned 15 boreholes in three low-income suburbs, Glen Norah, Mbare and Budiriro, contaminated with human fecal matter (*Escherichiacoli* bacteria) (Kafe, 2023). Such incidences have made household purification of water necessary.

The use of low-cost household purification technologies, like water filters, in Zimbabwe's urban areas is still limited. This explains the rampant cholera and typhoid outbreaks in urban areas. Aqua tablets are used as an alternative to low-cost technologies, but their distribution to needy households is not effective. Some of the low-income households resist using these tablets, citing that they produce an unpleasant smell in water. As such, they prefer boiling their water. However, with the prolonged power cuts in the country and the increasing cost of firewood, liquefied petrol gas and other alternative energy sources, most low-income households cannot afford to boil their drinking water regularly.

8.4 Policy Recommendations

A list of intertwined factors cause Zimbabwean urban water supply sector challenges, and this makes it difficult to come up with clear-cut policy strategies. However, policy options should address areas like economic recovery, improved urban governance, urban expansion patterns, private sector engagement, and technological advancements.

8.4.1 Economic Recovery

A protracted period of economic decline in Zimbabwe has negatively affected the functionality of different institutions, organizations and systems (mainly financial and social). Local municipalities have not been spared. Some are heavily dependent on government grants for their operational sustainability. Yet, the central government funds have been dwindling over time. Economic recovery is likely to benefit local municipalities in several ways. First, central government grants will increase, improving local municipalities' ability to finance water infrastructure investments. Second, an improved economy is associated with better wages. Thus, the brain drain is likely to be reduced. Local municipalities will be able to attract highly skilled personnel who can contribute to addressing water woes. A favorable economic and political situation improves the chances of local municipalities getting both local and international investors willing to invest in the development, refurbishment and renovation of water infrastructure. Moreover, in a thriving economy, most residents can pay their rates and taxes, improving municipal revenue. With improved municipal revenue, chances are high that service delivery will improve.

8.4.2 Improved Urban Governance

Good urban governance can improve the availability of funds for water infrastructure development and maintenance. Water finances are sometimes abused in local municipalities and are used for purposes other than water infrastructure development or water supply. Lack of separation of accounts makes it easier for some municipalities to divert water funds to other uses. Over the years, corruption issues have increased in local municipalities and within central government structures. Funds for service delivery improvement have disappeared without a trace, compounding water infrastructure financing challenges. For example, the study found that Harare municipality had raised a substantial amount of money for the development of Kunzvi Dam, a proposed new water source for Harare. When the central government made Kunzvi Dam a national project in the late 1990s, the funds Harare municipality had raised were transferred to the central government for administration purposes. However, the funds 'disappeared', and Kunzvi Dam is yet to be completed.

Lack of accountability and transparency in cities also affect water infrastructure development and maintenance. Urban residents are sometimes reluctant to settle their rates and taxes due to the alleged abuse of funds in municipalities. Lack of transparency in the generation of water bills has caused conflicts in suburbs in Harare. In Msasa Park, residents refused to settle service charges because they had gone for 3 years without receiving water supplies. The municipality argued that residents were supposed to pay for the maintenance of water infrastructure regardless of not receiving water supplies. In response, residents indicated that the municipality was free to remove their infrastructure because it was not benefitting them. This impasse negatively affected water revenue collection efficiency in the capital and had an impact on water service delivery.

8.4.3 Urban Expansion Patterns

Emerging research has shown that the growing demand for urban infrastructure in African cities is associated with their growth patterns (Gambe et al., 2023). City growth processes have resulted in rapid outward expansion. This growth pattern has increased the burden on local municipalities to provide critical infrastructure. Similar growth patterns are evident in Zimbabwean cities, mainly Harare, Bulawayo, Chitungwiza and Masvingo. Peri-urban developments have been pushing the cities' boundaries outward. While some of these developments have already been incorporated into the city boundaries, some are not. Outward expansions have a higher demand for infrastructure that municipalities are not able to provide. One way to address the growing need for water infrastructure is a policy that strictly encourages compact developments that take advantage of the existing infrastructure. Densification within the city's footprint is desirable, but without exceeding the carrying capacity of the existing infrastructure.

8.4.4 Private Sector Engagement

Private sector engagement can improve the financing of water infrastructure in Zimbabwean cities. The engagement can be in different forms, including co-financing arrangements, sharing water supply functions, and donations as part of corporate social responsibility.

Co-financing of water infrastructure is an arrangement that is popular in thriving and stable economies. However, economic volatility in Zimbabwe has made this difficult. The private sector is not forthcoming to participate in water infrastructure provision. The primary concerns are the failure to get lucrative returns and recoup capital. Encouraging private sector involvement in water provision entails legal guarantees from the central government and the local municipality to honor the financing agreement. Furthermore, warranties should be provided so that the private firm involved will be able to get a reasonable return and recoup its capital. These legal guarantees can soften the reluctance of the private sector to participate in water infrastructure provision. Central government can also award tax exemptions to firms in the private sector willing to partner with local municipalities in developing or refurbishing water infrastructure. This arrangement can be a lucrative incentive to the private sector.

Sharing water provision responsibilities is also a possible arrangement that encourages the private sector to participate in water infrastructure financing. In Chiredzi, a private firm named Hippo Valley Estates has developed a water treatment plant that supplies water to its employees living in the sugarcane estates. Apart from that, other residents in Chiredzi also get water supplies from this plant. The firm is responsible for maintaining the infrastructure and undertaking water purification and distribution processes. This arrangement can be replicated in other cities where the circumstances make it possible. For example, private companies that develop a significant number of housing units for their employees or sale should also be tasked with purifying and supplying water to the same units.

Local municipalities should also seek donations from private companies operating within their areas of jurisdiction. For example, cities or towns with mines operating within their boundaries can approach them for financial assistance to give back to the community. Community Share Ownership Trusts that were created in Zimbabwe some years back and which seem to be inactive, provided room for mining companies to give back to the communities in which they operated. The study recommends that this initiative be reviewed, revised and re-introduced to help local municipalities access corporate responsibility funds or donations from large corporations operating within their boundaries.

8.4.5 Technological Advancements

Although smart water metering has been adopted in Zimbabwean cities, its use has been mainly to measure the water consumption of the users and improve water revenue collection. Yet, these meters can be used in many ways that will benefit both the water managers and users. One of the factors that increased non-revenue water in Zimbabwean cities is the time taken to replace burst pipes. Most cities do not have smart tracking mechanisms to quickly alert the municipality of a burst pipe and its location. This is a gap that smart meters can plug and can help water utilities to respond swiftly to emergencies in the water supply system. Funds permitting, smart meters that monitor the water flows of the distribution systems are needed in Zimbabwean municipalities.

This study also recommends the use of smart meters in water audits. It is essential to accurately calculate the percentage of the water that reaches the end consumers after being pumped into the distribution systems. In this way, underground leakages are easily detected and attended to. These functions are useful to Zimbabwean municipalities struggling to detect and address underground leakages. Others are failing to accurately calculate city consumption patterns and, therefore, plan accordingly regarding supplies. However, it is crucial to recognize that some strides have been made in cities like Harare, Bulawayo, Kariba and Gweru, which have adopted geographic information systems to monitor different services. Burst water pipes are now easily located through the GIS monitoring techniques in Kariba. This system has significantly reduced non-revenue water in the city and water revenue lost through undetected water leaks.

The number of households in Zimbabwean cities using unprotected water sources is fast increasing owing to prolonged periods of water rationing. Unprotected wells have become a common water source in low-income suburbs and unplanned or partially planned settlements. Using low-cost household filters can reduce the outbreak of waterborne diseases such as cholera, typhoid and dysentery. Apart from low-cost household purification filters, an increased uptake of aqua tablets is recommended. Keeping them in health facilities and waiting for households to come and pick them up has proved ineffective. New public health programs that take these tablets to households need to be rolled out. However, the effectiveness of the water filters and aqua tablets depends on the attitude and behavior of the households. Behavior change programs are also required to promote hygiene when handling and preparing drinking water. Households should strive to consume clean water that has undergone at least some household-level purification, whether boiling, using aqua tablets or water purification filters.

8.5 Conclusion

This chapter explored the innovative financing mechanisms and technological advancements that can ease urban water shortages. The chapter addressed three research questions:

- What are the factors behind the water woes in Zimbabwean municipalities?
- What technological advancements have been made in the urban water supply sector?
- What policy strategies can be implemented to improve urban water supply in Zimbabwe's urban municipalities?

Local municipalities in Africa are struggling to raise the funds required to provide the necessary infrastructure to cater for the needs of the ever-growing urban population. In Zimbabwe, water infrastructure deficiencies have resulted in water scarcity in most cities and towns. The limited participation of the private sector in urban water supply has inhibited private financing of water infrastructure. Central and local government institutions dominate the water sector in Zimbabwe and operate on a cost-recovery basis. However, water revenue collection is not as efficient and effective. Revenue collection in most municipalities is affected by factors such as high percentages of non-revenue water and defective billing systems and the economic decline in the country. Urban residents cannot settle their rates and taxes in time due to affordability issues; consequently, local municipalities owed huge sums of money in outstanding rates and taxes, making water service delivery poor.

Synergies between local municipalities and local firms can help improve the water supply in urban centers. The model adopted by Harare City, which is working in partnership with local firms to replace imported chemicals with locally produced water purification chemicals, is commendable. The cost reduction emanating from this partnership is significant since Harare requires approximately eight chemicals to treat its heavily polluted water. Some of the chemicals are imported and need foreign currency, which is scarce. Similar synergies are recommended in a bid to find sustainable financing mechanisms that can improve investments in water infrastructure.

Other recommended strategies include improved governance, public–private partnerships in water supply with private companies involved in the purification and distribution of water, and technological advancements. Separation of accounts is recommended to improve transparency and accountability in local municipalities. Water revenue should be channeled towards improving the water supply as this motivates residents to settle their bills timeously and in full. Central government interference in the affairs of city councils should be limited and allow local municipalities to deliver services to their residents. Local municipalities need the space to make their strategic plans and implement them without interference.

The water provision model in Chiredzi can be replicated in other urban centers. Large private companies should be allowed to construct and operate water purification plants, supplying water to their employees while benefiting other residents in the process. However, there is a need for a strong partnership between firms offering such services and local municipalities (which must have a supervisory role) to protect the public health and welfare interests. Meanwhile, urban households should consider the uptake of low-cost household water filters that have become common in Africa's low-income communities. The filters can guarantee the consumption of clean and safe water, especially in unplanned and partially planned communities dependent on unprotected wells for household water supplies. Aqua tablets are still a feasible option but should be delivered to the households rather than waiting for them to collect at local health centers.

References

Bhanye, J. I., Mangara, F., Matamanda, A. R., & Kachena, L. (2023). *COVID-19 Lockdowns and the urban poor in Harare*. Springer.

Bulawayo 24 News. (2018). Gweru loses 57% treated water. *Bulawayo 24 News*, 5 June. https://bulawayo24.com/index-id-news-sc-national-byo-137255.html. Accessed 14 September 2023.

Dodman, D., Leck, H., Rusca, M., & Colenbrander, S. (2017). African urbanisation and urbanism: Implications for risk accumulation and reduction. *International Journal of Disaster Risk Reduction, 26*, 7–15. https://doi.org/10.1016/j.ijdrr.2017.06.029

Gambe, T. R. (2013). Water billing and service delivery management in Harare: A case of Upper Avenues and Msasa Park. In I. Chirisa (Ed.), *Contemporary rural and urban issues in Zimbabwe: Implications for policy and planning* (pp. 249–279). Academica Press.

Gambe, T. R. (2015). Prospects of prepaid smart water metering in Harare, Zimbabwe. *Afr J Sci Technol Innov Dev, 7*(4), 236–246. https://doi.org/10.1080/20421338.2015.1081762

Gambe, T. R. (2019). The gender dimensions of water poverty: Exploring water shortages in Chitungwiza. *J Poverty, 23*(2), 105–122. https://doi.org/10.1080/10875549.2018.1517399

Gambe, T. R., Turok, I., & Visagie, J. (2023). The trajectories of urbanisation in Southern Africa: A comparative analysis. *Habitat Int, 132*, Article 102747. https://doi.org/10.1016/j.habitatint.2023.102747

Kafe, E. (2023). Human waste detected in city boreholes. *The Sunday Mail*. 29 January. https://www.sundaymail.co.zw/human-waste-detected-in-city-boreholes. Accessed 18 March 2025.

Kusemachibi, M. (2022). EMA fines Masvingo City Council, Delta for environment pollution. *TellZim News*. 4 June. https://tellzim.com/ema-fines-masvingo-city-council-delta-for-environment-pollution/. Accessed 18 March 2024.

Madzimure, J. (2022). President launches borehole drilling scheme. *The Herald*. 24 March. https://www.herald.co.zw/president-launches-borehole-drilling-scheme/. Accessed 18 March 2025.

Maponga, G. (2022). Masvingo council probes fish deaths in two rivers. *The Herald*. 27 May. https://www.herald.co.zw/masvingo-council-probes-fish-deaths-intwo-rivers/. Accessed on 18 March 2024

Matamanda, A. R., Mphambukeli, T., & Chirisa, I. (2020). Applied systems analysis in water access for emerging human settlements: a case study of Hopley Farm, Harare, Zimbabwe. *Urban Water Journal, 17*(9), 763–773.

Mgwele, A., Bhanye, J. I., & Mocwagae, K. S. (2024). Water and sanitation for urban health: A gender perspective on impacts and coping strategies in Mangaung Metropolitan Municipality. In A. R. Matamanda & V. Nel (Eds.), *Sustainable development goals and urban health: Strides, challenges and way forward for poor neighborhoods* (pp. 49–67). Springer.

Moyo-Ndlovu, T. (2023). Bulawayo tops in water and sanitation services. *Chronicle*. 19 June. https://www.chronicle.co.zw/bulawayo-tops-in-water-and-sanitationservices/. Accessed 15 September 2023

Mpofu, M. (2022). All Bulawayo suburbs to go without water indefinitely. *Chronicles*. 30 December. https://www.chronicle.co.zw/all-bulawayo-suburbs-to-go-without-water-indefinitely/. Accessed 16 March 2025

Mubaiwa, S. (2023). Gweru runs dry. *Masvingo Mirror*. 11 September. https://masvingomirror.com/gweru-runs-dry/. Accessed 18 March 2025

Mukaro, B. (2023). Water woes giving us headaches-Masvingo Town Clerk. *Masvingo Mirror*. 4 April. https://masvingomirror.com/water-woes-giving-us-headachesmasvingo-town-clerk/. Accessed 4 April 2023

Mukuhlani, T., & Nyamupingidza, M. T. (2014). Water scarcity in communities, coping strategies and mitigation measures: The case of Bulawayo. *Journal of Sustainable Development, 7*(1), 144–160. https://doi.org/10.5539/jsd.v7n1p144

Muleya, M. (2023). Mutare City battles 60% non-revenue water. *The Mirror*. .

Musemwa, M. (2006). Disciplining a "dissident" city: Hydropolitics in the City of Bulawayo, Matabeleland, Zimbabwe, 1980-1994. *Journal of Southern African Studies, 32*(2), 239–254. https://doi.org/10.1080/03057070600656119

Musemwa, M. (2010). From "Sunshine City" to a landscape of disaster: The politics of water, sanitation and disease in Harare, Zimbabwe, 1980-2009. *J Dev Soc, 26*(2), 165–206. https://doi.org/10.1177/0169796X1002600202

Nel, E. L., & Berry, B. B. (1992). The problems of supplying water to third world cities: Bulawayo's water crisis. *Dev South Afr, 9*(4), 411–422. https://doi.org/10.1080/03768359208439648

Nhapi, I. (2009). The water situation in Harare, Zimbabwe: A policy and management problem. *Water Policy, 2*, 221–235. https://doi.org/10.2166/wp.2009.018

Rammile, S., Bhanye, J. I., Lehobo, T., & Hugo, M. (2024). Building resilient urban infrastructure and innovation for urban public health: A case study of Bloemfontein, South Africa. In A. R. Matamanda & V. Nel (Eds.), *Sustainable development goals and urban health: Strides, challenges and way forward for poor neighborhoods* (pp. 69–96). Springer.

World Bank. (2013). *Zimbabwe Urban Water Tariff Study*. The Zimbabwe Water Forum Policy Note 6. https://documents1.worldbank.org/curated/en/164151468196755260/pdf/103624-BRI-P126703-PUBLIC-ZWF-6-Tariff-Study-Sep-2013.pdf

Chapter 9
Political Instrumentalization of Urban Water Governance in Zimbabwe: A Water–Health Nexus

Abstract This chapter provides complex insights on how urban water governance challenges influence the water–health nexus because of the political instrumentalization of potable water. Concurrently, post-1980, the urban water infrastructural systems have increasingly depreciated, thereby leading to potable water challenges. In retrospect, this study highlights the water–health nexus because Zimbabwe has a history of frequent, severe cholera outbreaks that have resulted in at least 5000 deaths per outbreak on average; most people who died fell into the category of people living in poverty. The cholera pandemic is a stark illustration of the unique problems underpinning urban water service delivery, which fosters a favorable setting for a wide range of waterborne challenges. Furthermore, due to Zimbabwe's hyperinflationary economy, which obstructs low-income households from accessing alternative sources of potable water, people living in poverty are more prone to suffering from a wide range of water-related health diseases. As a result, the numerous health implications caused by fissures and cracks in the urban water governance structures must not be overlooked. The study also argues on how these health inferences extend beyond the generic waterborne diseases and reflect on how urban water challenges have created emotional and psychological stress, particularly for females who bear the household burden of ensuring the accessibility of potable water. In essence, the study makes the case that a complex range of factors, including groundwater contamination, infrastructural challenges, politicization of water bodies, and patterns of global climate change, are also unquestionably to blame for the urban water services sector's inability to deliver potable water efficiently and effectively. Thus, as the chapter's conclusion underscores, water is an essential element of sustainable development and ensuring human health and well-being. In conclusion, this study illuminates the pervasive influence of political interests on urban water governance in Zimbabwe and its detrimental consequences for public health and sustainable development.

Keywords Health · Potable water · Urban water supply · Sustainability · Water · Health nexus

A. R. Matamanda et al., *Urban Water Governance in Postcolonial Zimbabwe*, Local and Urban Governance, https://doi.org/10.1007/978-3-031-96973-7_9

9.1 Introduction

Insufficient equipment and infrastructure pose serious barriers to the delivery of sustainable services in urban areas (Rammile et al., 2024). Zimbabwe is characterized by the sharp fall in public sector investment in new urban service delivery infrastructure and equipment in the last forty years. Zimbabwe's urban areas suffer from poor service delivery due to several issues, including infrequent water supply, burst water and sewer pipes, faecal contamination of major water sources, deteriorating road networks, broken traffic lights, uncollected refuse, unfinished capital projects, and protests against poor service delivery (Bhanye et al., 2023). The Report of the Auditor-General (2019) on local authorities presented to the Parliament of Zimbabwe for the 2018 financial year painted a grim picture of inadequate service provision in the country's cities, towns, and smaller urban settings. It arguably depicts a pitiful picture of service delivery characterized by infrastructural paralysis, massive strain, poor and nonexistent water provision, intermittent refuse collection and dilapidated service delivery infrastructure.

In urban Harare, the water–health nexus manifests as a critical issue shaped by various socioeconomic and environmental factors. With a rapidly growing population and strained infrastructure, access to clean water and sanitation facilities becomes increasingly challenging (Kasimba et al., 2023). The inadequate provision of safe drinking water and proper sanitation services further heightens the risk of waterborne diseases, including cholera and typhoid, which pose significant threats to public health. Moreover, Samhokore (2022) argued that frequent water shortages and interruptions further exacerbate these challenges, forcing residents to resort to unsafe water sources or engage in risky hygiene practices, amplifying the likelihood of disease transmission. Furthermore, the water–health nexus in urban Harare underscores systemic disparities in access to basic services, disproportionately affecting marginalized communities and low-income households who bear the brunt of inadequate water infrastructure, resulting in health risks and reduced quality of life (Marumahoko et al., 2020). This study argues that addressing the water–health nexus in urban Harare requires holistic approaches that address not only infrastructural deficiencies but also socioeconomic inequalities and political issues to ensure equitable access to safe water and improved public health outcomes for all residents.

This chapter aims to investigate the political instrumentalization of urban water governance in Zimbabwe within the context of the water–health nexus. By examining the interplay between political actors, water management policies, and public health outcomes in urban areas of Zimbabwe, the chapter uncovers the extent to which political dynamics influence decision-making processes, resource allocation, and infrastructure development in the provision of clean and accessible water. The chapter sheds light on the underlying mechanisms and implications of political instrumentalization on urban water governance, using an interdisciplinary approach that integrates perspectives from political science, public health, and environmental studies, with a focus on understanding its impacts on public health outcomes and broader socioeconomic development in Zimbabwe. The chapter addresses the following research questions:

- How does political instrumentalization influence urban water governance in Zimbabwe?
- What are the implications of the political instrumentalization on the water–health nexus?
- What sustainable policy interventions can help improve urban water management and mitigate health risks?

The research questions underscore the urgent need for integrated governance solutions that prioritize public health, infrastructure investment, and equitable water access to advance sustainable development in Zimbabwe, as discussed in this chapter.

9.2 Literature Review

This section provides a comprehensive overview of the water–health nexus. The chapter further provides an in-depth linkage between the water–health nexus and the SDGs, a global perspective on water quality and urban water governance and health implications in Zimbabwe, and the cholera outbreaks in Zimbabwe, giving relevance to the water–health nexus. In Zimbabwe, urban water governance has become entangled in political agendas, leading to complex dynamics that significantly affect public health. This conceptual framework explores the intricate relationship between political instrumentalization, urban water governance, and public health within the Zimbabwean context (Fig. 9.1).

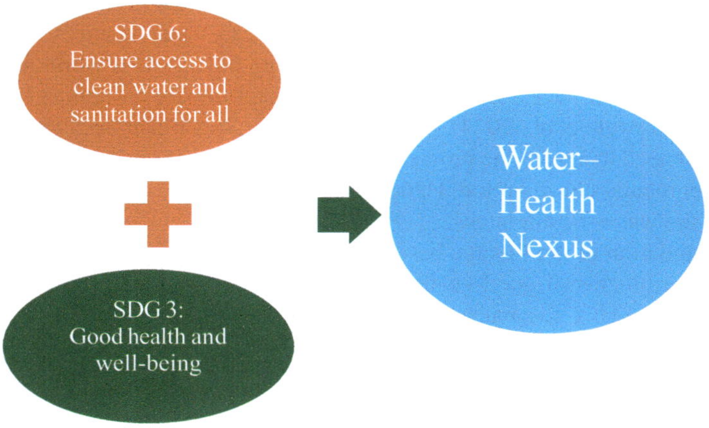

Fig. 9.1 Water–health nexus and sustainable development goals. (Source: Authors, 2024)

Musemwa (2021) argued that political instrumentalization refers to the manipulation of water governance systems for political gain or control. This study argues that, in Zimbabwe, political actors often use access to water resources and infrastructure to exert power, influence public opinion, and maintain political stability. Urban water governance encompasses the institutions, policies, and processes that govern the management and provision of water services in urban areas (Harvey, 2022; Matsengarwodzi, 2023). In Zimbabwe, urban water governance faces numerous challenges, including corruption, inefficiency, and inadequate infrastructure (Afrobarometer, 2021). As a result, the political instrumentalization further complicates governance structures, leading to disparities in access to clean water and across different urban centers. The water–health nexus examines the interconnectedness between water quality, sanitation, and public health outcomes. In Zimbabwe, poor urban water governance exacerbates health risks, including waterborne diseases such as cholera and typhoid. Katekar and Deshmukh (2021) highlighted that inadequate access to clean water and sanitation facilities contributes to high morbidity and mortality rates, particularly among vulnerable populations such as children, the elderly and low-income households.

9.2.1 Water–Health Nexus and Sustainable Development Goals

The SDGs provide a blueprint of 17 goals towards a global partnership to achieve sustainable development. The blueprint recognizes the urgent need for collaborative action to alleviate poverty, reduce inequalities, and act against the devasting effects of climate change. It is worth noting that these SDGs are interrelated, and one goal influences another, for instance, SDG 6 is linked to SDG 1 (No poverty), SDG 3 (Good health and well-being), and SDG 5 (Gender inequalities). Relatively, the water–health nexus in the context of SDGs is linked to SDG 3 and SDG 6, which condense the linkage between the provision of clean water services and human health and well-being. Although there is an existing linkage, it is worth noting that water resources can have a detrimental or beneficial impact on human health and well-being (Mgwele et al., 2024). Equally important, this illustrates that there is an exponential need to ensure the provision of water that is safe and healthy for human and environmental consumption. Hence, Fig. 9.1 depicts the linkage of SDG 3 and SDG 6 to the water–health nexus.

The foundation of the linkage between SDG 3 and SDG 6 postulates that one of the key determinants of good health and well-being is access to safe and healthy water services. This resonates with Target 6.3 of SDG 6 which provides the following:

> *By 2030, improve water quality by reducing pollution, eliminating dumping and minimizing release of hazardous chemicals and materials, halving the proportion of untreated wastewater and substantially increasing recycling and safe reuse globally.*

On the other hand, Target 3.9 of SDG 3 provides that:

> *By 2030, substantially reduce the number of deaths and illnesses from hazardous chemicals and air, water and soil pollution and contamination* (Advocates for International Development, 2019).

In support, empirical studies show that in sub-Saharan Africa, one in every ten children under five years old dies due to diarrhea (World Health Organization [WHO], 2020; UNICEF, 2023). According to the WHO (2020), diarrheal disease is the second leading cause of death in children under five years old and more than 370,000 children died globally in 2019 due to diarrhea. Several studies (Amadu et al., 2023; Malebatja & Mokgatle, 2022) conceded and highlighted that diarrhea is mostly linked due to inadequate water quality, poor sanitation facilities and hygiene. Similarly, Hall et al. (2020) conducted a study on water and health interlinkages in remote Indigenous communities in Australia. The study showed that most populations in these remote Indigenous communities in Australia were experiencing a burden of diseases. The study further revealed that most of the diseases were waterborne diseases or were hygiene-related, which included diarrhea and endemic trachoma eye infections due to unsafe water services in the area (Hall et al. 2020). However, evidence from the 2020 WHO report giving context to the COVID-19 pandemic, emphasized the importance of safe water, sanitation and hygiene as a precautionary measure that was used to protect human health to reduce the transmission of the COVID-19 virus (Bhanye & Bhanye, 2023). In essence, the importance of water cannot be over-emphasized in promoting good health and well-being to citizens. Hence, it is imperative to view the water–health nexus as a holistic approach that should be included in global, national, and local policies to give effect to the active realization of the humans' right to water (SDG 6) and health care (SDG 3). Thus, this further illustrates the exponential need to invest in the maintenance and upgrading of water treatment facilities, simultaneously decreasing cross-contamination and waterborne diseases.

One of the fundamental elements of the water–health nexus is the convergence of water quality and human health (Confalonieri & Schuster-Wallace, 2011). In context, water quality is one of the main twenty-first-century challenges that countries and societies are currently facing. Water quality has deteriorated due to pollution in the major rivers of Latin America, Africa, and Asia. Notably, low water quality poses a detrimental impact on human health, food production levels and economic growth. Evidence from the International Initiative on Water Quality revealed that worldwide, one in nine people use drinking water from unsafe and unimproved water sources. This is because 90% of sewage in developing countries is discharged untreated directly to water bodies (International Initiative on Water Quality).

However, most of these countries do not have modern and industrial water treatment systems owing to the complexity and affordability of these systems. As a pointer, water treatment in developed nations such as the United States utilizes complex steps that include water chemistry and engineering elements (Carmody, 2020). This includes steps such as coagulation, sedimentation, filtration, and disinfection. These measures include a holistic approach to water treatment to ensure the provision of quality water. Unfortunately, constructing and operating such water treatment systems in developing countries is improbable due to financial constraints, infrastructural backlogs, and institutional capacities (Carmody, 2020). Consequently, without adequate systems for water treatment, most developing countries turn to alternative solutions. It is against this background that the chapter acknowledges that some of the most common water treatment solutions used in developing countries include chlorination, solar water disinfection (SODIS), ceramic and bio-sand water filters, and natural groundwater.

As previously mentioned, the lack of financial and technological resources impedes the implementation of modern drinking water treatment plants in these regions. First, it is worth noting that researchers such as Cowie et al. (2020) agreed that the SODIS system is cost-effective in low-to-medium-income countries. Although there are benefits to using the SODIS system, the major disadvantage is the long radiation exposure required to ensure satisfactory water disinfection levels (6 hours on a clear sunny day, and 48 hours under cloudy conditions) (Cowie et al., 2020). Additionally, the containers or bottles used for radiation may be unclean and this can still pose a health risk to consumers. In principle, such systems support the first target of SDG 6, which is to "achieve access to safe and affordable drinking water." However, there is still a need to take cognizance of the affordability element considering the limited financial resources available to ensure sustainable provision of clean water.

Second, the chlorination system is used in most countries to ensure the provision of potable water (Zerbo et al., 2021). Chlorination entails the process of adding chlorine chemicals to kill bacteria, parasites, and viruses (Centers for Disease Control and Prevention, 2020). Briefly, chlorination is said to be one of the major disinfectants used in public water systems. However, the efficiency of chlorine in human consumption is still questionable. The Centers for Disease Control and Prevention (2020) denoted that using or drinking water with small amounts of chlorine does not cause harmful health effects but rather protects against waterborne disease outbreaks. The only key issue is that the taste and smell problems may arise depending upon the water quality and amount of chlorine in the water. Nonetheless, chlorination provides viability for water quality at reduced costs for low- and medium-income countries.

Third, as previously outlined, a ceramic filter entails the process of water flowing through a filter with tiny pores, preventing contaminants from passing through. Studies have shown that ceramic filters can be three to six times more cost-effective than a centralized water system in place (Carmody, 2020). Unfortunately, although this system is more cost-effective, these filters do not completely remove viruses or bacteria from the water (Yang et al., 2020). This means that the water can still be highly unfit for human consumption with major health risk implications.

Lastly, groundwater is pivotal to fight against poverty, water, and food security. The United Nations World Water Development Report (2022) highlighted that approximately 50% of the world's urban population relies on groundwater sources. Unfortunately, bearing in mind the infrastructural backlogs, the water is polluted, mismanaged, and undervalued. If properly managed, Harvey (2022) asserted that groundwater resources are enough to transform agriculture and increase access to safe drinking water. However, the lack of investment in these resources is progressively slow to meet water demands. Groundwater is often polluted due to human activity and industrialization. Additionally, groundwater sources are not disinfected at all, which replicates a high possibility of cross-contamination from the pipes through which the water run.

It is important to note that although these treatments are effective in removing some contaminants, they also have their drawbacks and do not fully treat water, which still poses a major health hazard to citizens. Carmody (2020) accentuates that drinking unsafe water can lead to a variety of health risks in humans. Moreso, one of the contaminants most found in water are pathogenic microorganisms which lead to outbreaks of waterborne diseases such as typhoid fever, cholera, and giardiasis. Carmody (2020) further stressed that such illnesses can be mortal to vulnerable populations, who may also lack access to medical healthcare services. On that note, 11 counties in Eastern and Southern Africa are currently battling with one of the worst cholera outbreaks to hit the region in years (UNICEF, 2023). These countries include Mozambique, Malawi, Somalia, Kenya, Ethiopia, Burundi, Tanzania, South Africa, South Sudan, Zimbabwe, and Zambia. Resultantly, out of 11 countries, Malawi, Mozambique, Zimbabwe, and South Sudan have been noted as "priority" countries due to the intensity of the cholera outbreaks in these countries (UNICEF, 2023). To complicate matters even further, in March 2023, Malawi was hit by Tropical Cyclone Freddy, which displaced about 600 people (UNICEF, 2023). Most of them are still living in temporary shelters with compromised water and sanitation facilities, posing a high risk of waterborne diseases, including cholera.

9.3 Urban Water Governance and Health Implications in Zimbabwe

Mutandwa and Vyas-Doorgapersad (2023) highlighted that in rapidly urbanizing areas, the provision of clean and accessible water services is essential to prevent the spread of diseases and maintain public health. Consequently, proper wastewater treatment and sewage disposal systems are equally important. It is noted that Zimbabwe's main water sources include surface water from the river, treated with different methods such as an ultrafiltration system, groundwater and water from the government's supply (Hoko et al., 2024). Although Zimbabwe upholds a constitutional mandate under section 77 of the COZ, or the government takes reasonable measures to ensure the provision of potable water services, there have been inconsistencies and crises in the provision of potable water services in Zimbabwe.

Additionally, many parts of Zimbabwe continued to face water issues, with some places in Harare going without safe water for years, and residents turning to potentially contaminated wells and boreholes (Maramura, 2017). Currently, Harare's greater metropolitan area has about 4.5 million people, more than half of whom have no access to clean water and are at risk of waterborne diseases such as cholera and typhoid (Human Rights Watch, 2023). Furthermore, the Human Rights Watch reports (2021, 2023) revealed that neighboring towns like Chitungwiza, Ruwa, Epworth, and Norton rely on water from Harare's water system, which is reportedly contaminated with algae and toxic substances linked to incidents of waterborne diseases in those locations. Predictably, due to Zimbabwe's hyperinflationary economy, low-income households are also obstructed from accessing sustainable alternatives to potable water and are more prone to suffering from a wide range of water-related health diseases (Zimbabwe Statistics, 2022). Those who have no access to tap water rely on boreholes for all their water, which is said to be unclean. In Harare's low-density suburbs in the north and east of the city, more affluent families have devised alternatives (Maramura, 2017), including drilling safe boreholes and using bottled and filtered water, which costs Zim $86 (US$1) for a one-liter bottle (Bhanye et al., 2023).

According to Human Rights Watch (2021), corruption and mismanagement are the key factors affecting the allocation of water resources, their quality, and distribution. However, the water crisis has subsequently prevented the formation of necessities such as flushing systems in most regions in Zimbabwe. Consequently, most households in Zimbabwe are forced to utilize buckets and bushes (Maramura, 2023). Under such arrangements, the lack of proper sewage disposal systems, hot temperatures and accidental spilling of these buckets cause an immense health risk such as worm infestations and gastrointestinal disorders. The Water Research Commission (2020) avowed and highlighted that bucket systems are a violation of human dignity and exposure to health risks. This further illustrates the health risks that the Zimbabwean government exposes its citizens to due to inadequate provision of water services and violating section 76 of COZ, which guarantees the right to health to its citizens.

Moreover, a study by Samhokore (2022) investigated the impact and coping means of water, sanitation and hygiene in the Caledonia informal settlement in Harare, Zimbabwe. According to the findings of the study, the vast majority of populations in Caledonia experience diarrhea after drinking water. Not only in Zimbabwe but the health impact of unsafe water and the associated risk of contamination from unmanaged sewage is estimated to cause around 60% of deaths from diarrhea globally. Adding to these health woes, the water issues in Caledonia cause sexual and reproductive challenges, depression, and psychological stress on women, who bear the responsibility for household water use (Samhokore, 2022). Hence, it is critical to emphasize how these health interferences extend beyond generic waterborne diseases and also reflect on how urban water challenges exacerbate gender inequalities and stagnate social progress for women.

9.3.1 Cholera Outbreak in Zimbabwe

In retrospect, this chapter highlights the water–health nexus because Zimbabwe has a history of prevalent, severe typhoid and cholera outbreaks that have resulted in at least 5000 deaths on average during every outbreak. In 2008, Zimbabwe experienced a spiraling political and economic crisis, and as a result, most public services collapsed (Matsengarwodzi, 2023). On that note, fresh water supplies were not consistent, leading most women and children to scavenge unsafe water and wells. The conditions of these wells were poor and contaminated by unmanaged raw sewerage systems. Following that within the same year, almost 100,000 cholera outbreaks were reported, and 400,000 official deaths were reported (Matsengarwodzi, 2023). Surprisingly, that was the most tremendous cholera outbreak in Africa and the nation's history. Despite these misfortunes, exactly 10 years after, in 2018, another cholera outbreak hit Zimbabwe, with over 10,370 cases being reported. The 2018 outbreak was noted as "better" compared to the devasting effects of the 2008 cholera outbreak. On that note, under section 77 of the COZ, "every person has the right to safe, clean, and potable water". These cholera outbreaks arguably reflect the violation of the human rights of millions of Zimbabwean citizens. Tentatively, the government should invest in low-cost water equipment and distribution systems to uphold the right of millions of Zimbabweans to potable water services (UNICEF, 2023).

Over and above that, the water crisis in Harare has affected people's rights to water and sanitation as well as other related rights, including the rights to life, food, and health. Machingura and Chandiwana (2023) added that, coupled with the devasting water issues of Zimbabwe, the country has experienced constant cholera and typhoid outbreaks in Harare, which have been associated with municipal water shortages and increased use of contaminated boreholes and shallow wells. The combination of poor water quality and sanitation and urban overcrowding continues to be a persistent driver of seasonal outbreaks of waterborne diseases in Zimbabwe. Several factors have further contributed to Harare's severe water problems, including economic decay, perennial droughts affecting Lake Chivero, which is dammed to supply Harare with water, the lack of maintenance of the old water infrastructure, the inability to procure the necessary chemicals to treat water sources, political struggles between the central government under the ruling party and the opposition-controlled city council, and corruption exacerbating the water crisis in Zimbabwe (Matsengarwodzi, 2023).

Moreover, the government has taken some positive steps to address the crisis, including setting aside US$5.3 million for a critical Harare water treatment project to supply water in most areas in Zimbabwe (Human Rights Watch, 2023). This relates to the shortage of treatment chemicals such as chlorine and aluminum used in water treatment and water infrastructure that has not been maintained and upgraded. However, there have been delays in disbursing the funds, exposing residents to further health risks. Consequently, the lingering water crisis has affected the rights of Zimbabweans under section 77 of the COZ, which provides for the right to safe, clean, and potable water. The Human Rights Watch (2023) highlighted that "Harare's long unresolved water crisis is a ticking bomb of magnified health risks that enforces residents to seek alternative, often unsafe water sources". The cholera pandemic is a stark illustration of the unique problems underpinning urban water service delivery, which fosters a favorable setting for a wide range of waterborne diseases in Zimbabwe.

9.3.2 COVID-19 Pandemic as a Catalyst for Political Instrumentalization of Urban Water Governance in Zimbabwe

This section specifically focuses on how the COVID-19 pandemic has been and was used as a tool for political agendas within the framework of urban water governance in Zimbabwe. The study confirms how the COVID-19 pandemic served as a pivotal moment for the political instrumentalization of urban water governance in Zimbabwe, where access to clean water has long been a contentious issue. As the pandemic unfolded, the importance of adequate water supply for sanitation and hygiene became even more pronounced, with handwashing touted as one of the primary measures to prevent the spread of the virus (Hall et al., 2020). However, the Zimbabwean government's response to ensuring equitable access to water has been marred by political considerations, exacerbating existing inequalities. Political leaders leveraged the pandemic to reinforce their authority, using water access as a tool for garnering support or punishing dissent. Consequently, this led to uneven distribution of water resources, with marginalized communities bearing the brunt of inadequate access, further deepening social disparities (WHO, 2023).

Moreover, the pandemic laid bare the systemic weaknesses and corruption within Zimbabwe's urban water governance structures. Political elites were also argued to be exploiting the crisis to siphon off public funds meant for water infrastructure development, leading to rampant mismanagement and dilapidated water systems in many urban areas. Meanwhile, efforts to address these issues continue to be hindered by political interference (Afrobarometer, 2022), with public authorities prioritizing short-term political gains over long-term sustainable solutions that benefit the public. As a result, the nexus between COVID-19, water governance, and politics further exacerbated a vicious cycle of neglect and exploitation, leaving ordinary Zimbabweans vulnerable to both the health risks of the pandemic and the systemic failures of their governance systems.

9.4 Discussion and Analysis

The discussion section delves into the intricate interplay between political dynamics and urban water governance in Zimbabwe, elucidating the manifold implications for public health, equitable resource distribution, and sustainable development as discussed below.

9.4.1 Impact on Water Infrastructure and Service Delivery

The politicization of water governance had a direct impact on water infrastructure development and service delivery, when resources are directed towards areas with strong political influence within the ruling party's constituencies (Taonameso et al., 2022), thereby leaving marginalized communities underserved. There is no doubt that this skewed allocation exacerbates inequalities in access to clean water and sanitation facilities, perpetuating socioeconomic disparities and compromising public health outcomes. A respondent from an Afrobarometer survey highlighted that:

> *The impact of political instrumentalization on water infrastructure development and service delivery cannot be understated. Daily, marginalized communities suffer disproportionately from inadequate access to clean water and sanitation facilities, perpetuating socioeconomic inequalities and compromising public health outcomes.*

Human Rights Watch (2020) reported that while 84% of Zimbabweans had access to safe drinking water in 1988, that proportion had shrunk to 72% by 2000 and 64% by 2017. More than four million people in the greater Harare metropolitan area have no household access to safe water for drinking. The survey findings from the Afrobarometer (2022) as depicted in the image above also reflected on how 51% of the respondents repeatedly or always had to cope without water. Zimbabwe Statistics (2022) also highlighted that from the 51% respondents, low-income households are mostly obstructed from accessing sustainable alternatives to potable water and are more prone to suffering from a wide range of water-related health diseases.

9.4.2 Political Instrumentalization of Urban Water Governance

Zimbabwe is one of the failed states of Africa due to the violence and power monopoly wielded by the ruling party (ZANU–PF) which has been in power since 1980 (Taonameso et al., 2022). In 2007, Zimbabwe ranked fourth on the Failed States Index Scores of the Fund for Peace, behind only Sudan, Iraq and Somalia, and at the head of the indicators of "*mounting demographic pressure*" and "*sharp* and/or *severe economic decline*" (Marumahoko et al., 2020). Zimbabwe also held parliamentary elections every five years in 1980, 1985, 1990, 1995, 2000, 2005, 2008, 2013, 2018, and 2023, all of which were won by the ruling party ZANU–PF and all of which triggered accusations of fraud, violence and intimidation. Machingura and Chandiwana (2023) asserted that the country's economy and humanitarian situation were deteriorating daily. This is augmented by survey findings from the Afrobarometer (2024), which highlighted that 72% of citizens say Zimbabwe is going in the wrong direction.

The political instrumentalization of urban water governance in Zimbabwe is a concerning trend, where political interests of the ruling party overshadow the fundamental goal of providing safe and reliable water access to all citizens. The instrumentalization manifests through various channels, including resource allocation, decision-making processes, and the prioritization of certain communities over others based on political affiliations. This leads to skewed resource allocation and decision-making processes, thereby undermining the effectiveness of water management efforts. Political interests often overshadow the primary goal of providing safe and reliable water access to citizens. This also impedes the citizen's human right to water as stipulated in the COZ.

9.4.3 Challenges in Addressing the Water–Health Nexus

The nexus between water governance and public health emerges as a critical concern for sustainable development (Confalonieri & Schuster-Wallace, 2011). However, addressing the nexus between water governance and public health is a daunting challenge in Zimbabwe due to poor governance practices which exacerbate waterborne diseases such as cholera and typhoid, posing significant risks to public health. The politicization of water management further impedes efforts to effectively address these public health challenges in a country that is already classified as a failed state (Maramura, 2023; Taonameso et al., 2022). Furthermore, a lack of accountability and transparency hinders the implementation of sustainable solutions, thereby perpetuating a cycle of water insecurity and health risks.

9.4.4 Community Perspectives and Participation

Toriro (2024) asserted that residents had first-hand experience of the consequences of political instrumentalization that mostly included erratic water supply, contamination issues, and inadequate sanitation facilities. However, in Zimbabwe, their voices are frequently marginalized in decision-making processes, thereby limiting their ability to influence policy outcomes and hold authorities accountable. Survey findings from Afrobarometer, which was based on the citizens' views on the government's performance from 1999 to 2017, indicated that community participation was limited due to the political instrumentalization. On average, 50% to 60% of the respondents indicated that their perspectives were not acknowledged in the decision-making processes. Machingura and Chandiwana (2023) supported that community perspectives offered valuable insights into the impact of political instrumentalization on urban water governance.

9.4.5 *Implications for Sustainable Development*

Addressing the political instrumentalization of urban water governance is crucial for achieving the SDGs in Zimbabwe. Access to clean water and sanitation is fundamental to poverty reduction, public health improvement, and economic growth (UNICEF, 2023; WHO, 2020). Therefore, by prioritizing the needs of citizens over political interests, policymakers can create an enabling environment for equitable water distribution, enhancing public health outcomes, and laying a solid foundation for long-term socioeconomic development.

9.5 Recommendations

To address the pervasive politicization of urban water governance in Zimbabwe, it is imperative to prioritize depoliticization measures. This entails establishing independent regulatory bodies to oversee resource allocation and infrastructure development, ensuring transparent decision-making processes, and reducing political interference. By fostering a more impartial governance framework, authorities can better uphold the principles of fairness and equity in water service delivery and avert the political–administration dichotomy. Despite the challenges identified, there are opportunities for reforming urban water governance in Zimbabwe. Strengthening institutional frameworks, enhancing transparency and accountability mechanisms, and promoting community participation are essential steps towards depoliticizing water management. Civil society engagement, supported by international partnerships and donor initiatives, can play a pivotal role in advocating for policy reforms and fostering inclusive decision-making processes. Community participation is key to fostering inclusive and responsive water governance.

Empowering local communities through participatory decision-making mechanisms, such as participatory budgeting and community-based monitoring, also has the potential to enhance accountability and ensure that the voices of marginalized groups are heard. By involving citizens in the planning, implementation, and monitoring of water projects, authorities can better address the diverse needs and priorities of urban residents. Investment in water infrastructure development must also be prioritized, particularly in underserved urban areas and informal settlements. Allocating resources based on need rather than political considerations is essential to improving access to clean water and sanitation facilities for all citizens, thereby enabling a resilient and sustainable urban water governance system in Zimbabwe, one that prioritizes the well-being of its citizens and supports long-term socioeconomic development.

9.6 Conclusion

This chapter has examined how the political instrumentalization of urban water governance in Zimbabwe has contributed to deteriorating infrastructure, inequitable water access, and severe public health consequences, particularly within the water–health nexus. These persistent governance failures, coupled with economic instability, have exacerbated several waterborne diseases that disproportionately affect low-income households. Therefore, in line with SDG 6, addressing these challenges requires sustainable policy interventions that depoliticize water management, strengthen institutional capacity, and promote infrastructure investment. An integrated governance approach that prioritizes public health, equitable water access, and climate-resilient infrastructure is also necessary for reducing health risks and improving sustainable urban water management in Zimbabwe. In conclusion, this chapter illuminated the pervasive influence of political interests on urban water governance in Zimbabwe and its detrimental consequences for public health and sustainable development. The findings highlighted the urgent need for comprehensive reforms to depoliticize water management, enhance transparency, and promote inclusive decision-making processes. By prioritizing the equitable distribution of resources and prioritizing the needs of marginalised communities, policymakers can mitigate the impacts of political instrumentalization and pave the way for a more resilient and inclusive water infrastructure system.

References

Advocates for International Development. (2019). SDG 3: good health and well-being. *Advocates for International Development*. https://www.a4id.org/wp-content/uploads/2019/10/SDG-Legal-Guide_Chapter-3_V2.pdf

Afrobarometer. (2021). Summary of results: Afrobarometer Round 8 survey in Zimbabwe, 2021. https://www.afrobarometer.org/publication/zimbabwe-round-8-summary-of-results/

Afrobarometer. (2022). Summary of results: Afrobarometer Round 9 survey in Zimbabwe, 2022. https://www.afrobarometer.org/publication/zimbabwe-round-9-summary-of-results/

Afrobarometer. (2024). Summary of results: Afrobarometer Round 10 survey in Zimbabwe, 2024. https://www.afrobarometer.org/publication/zimbabwe-round-10-summary-of-results/

Amadu, I., Seidu, A.-A., Agyemang, K. K., Arthur-Holmes, F., Duku, E., Salifu, I., Bolarinwa, O. A., Hagan, J. E., Jr., & Ahinkorah, B. O. (2023). Joint effect of water and sanitation practices on childhood diarrhoea in sub-Saharan Africa. *PLoS One, 18*(5), Article e0283826. https://doi.org/10.1371/journal.pone.0283826

Bhanye, J., & Bhanye, A. (2023). A blessing or a curse? The role of social media during the COVID-19 pandemic in Africa. In R. Baikady, S. M. Sajid, J. Przeperski, V. Nadesan, M. R. Islam, & J. Gao (Eds.), *The Palgrave handbook of global social problems* (pp. 1–23). Palgrave Macmillan. https://doi.org/10.1007/978-3-030-68127-2_367-1

Bhanye, J. I., Mangara, F., Matamanda, A. R., & Kachena, L. (2023). *COVID-19 lockdowns and the urban poor in Harare, Zimbabwe: Emerging perspectives and the morphing of a sustainable urban future*. Springer.

Carmody J. (2020). Water treatment in developing countries, Office of Sustainability—Student Blog., 17 April. https://usfblogs.usfca.edu/sustainability/2020/04/17/water-treatment-in-developing-countries/. Accessed 18 March 2025.

Centers for Disease Control and Prevention. (2020). About water disinfection with chlorine and chloramine. https://www.cdc.gov/drinking-water/about/about-water-disinfection-with-chlorine-and-chloramine.html

Confalonieri, U. E., & Schuster-Wallace, C. J. (2011). Data integration at the water–health nexus. *Current Opinion in Environmental Sustainability, 3*(6), 512–516. https://doi.org/10.1016/j.cosust.2011.10.003

Cowie, B. E., Porley, V., & Robertson, N. (2020). Solar disinfection (SODIS) provides a much-underexploited opportunity for researchers in photocatalytic water treatment (PWT). *ACS Catalysis, 10*(20), 11779–11782. https://doi.org/10.1021/acscatal.0c03325

Hall, N. L., Creamer, S., Anders, W., Slatyer, A., & Hill, P. S. (2020). Water and health interlinkages of the sustainable development goals in remote indigenous Australia. *NPJ Clean Water, 3*, Article 10. https://doi.org/10.1038/s41545-020-0060-z

Harvey, F. (2022). Better use of groundwater could transform Africa, research says. *The Guardian*. 21 March. https://www.theguardian.com/environment/2022/mar/21/better-use-of-groundwater-could-transform-africa-research-says

Hoko, Z., Mapenzauswa, C. F., Toto, N. T., Kerith, M., & Nhapi, I. (2024). Analysis of operation and maintenance arrangements for water supply in urban areas in Zimbabwe. *Sustainable Water Resources Management, 10*, Article 27. https://doi.org/10.1007/s40899-023-01000-3

Human Rights Watch. (2020). World report 2020. Annual review of human rights around the globe. https://www.hrw.org/world-report/2020

Human Rights Watch. (2021). World report 2021. Annual review of human rights around the globe. https://www.hrw.org/world-report/2021

Human Rights Watch. (2023). World report 2023. Zimbabwe: Events of 2022. https://www.hrw.org/world-report/2023/country-chapters/zimbabwe

Kasimba, R., Choguya, N. Z., & Muqayi, S. (2023). Women and solid waste management in high-density areas: The case of Hatcliffe, Harare, Zimbabwe. *African Journal of Gender, Society & Development, 12*(1) https://hdl.handle.net/10520/ejc-aa_jgida1_v12_n1_a10, 209–228.

Katekar, V. P., & Deshmukh, S. S. (2021). Energy–drinking water–health nexus in developing countries. In M. Asif (Ed.), *Energy and environmental security in developing countries. Advanced sciences and technologies for security applications* (pp. 411–445). Springer. https://doi.org/10.1007/978-3-030-63654-8_17

Machingura, F., & Chandiwana, I. (2023). Hate speech in the context of water crises in Hatcliffe, Zimbabwe. In F. Machingura & N. Sande (Eds.), *Politics, religion and hate speech in Zimbabwe* (pp. 191–206). African Books Collective. https://doi.org/10.2307/jj.12949080.14

Malebatja, M. F., & Mokgatle, M. M. (2022). Diarrhea among children aged 5 years and microbial drinking water quality compliance: Trends analysis study in South Africa (2008–2018). *International Journal of Environmental Research and Public Health, 20*(1), Article 598. https://doi.org/10.3390/ijerph20010598

Maramura, T. C. (2017). The efficacy of prepaid water meters for potable water service provision in the Harare City Council, Zimbabwe. (Unpublished doctoral dissertation, North West University, Potchefstroom, South Africa). http://hdl.handle.net/10394/31633.

Maramura, T. C. (2023). Exploring the empirical efficacy of prepaid water meters in a failed state: a case of Harare in Zimbabwe. *Water Conservation & Management, 7*(2), 89–96. https://doi.org/10.26480/wcm.02.2023.89.96

Marumahoko, S., Afolabi, O. S., Sadie, Y., & Nhede, N. T. (2020). Governance and urban service delivery in Zimbabwe. *Strategic Review for Southern Africa, 42*(1), 41–68. http://hdl.handle.net/2263/81400

Matsengarwodzi, D. (2023). After three cholera outbreaks, Zimbabwe is on high alert to curtail cholera. https://www.gavi.org/vaccineswork/after-three-outbreaks-zimbabwe-high-alert-curtail-cholera

Mgwele, A., Bhanye, J. I., & Mocwagae, K. S. (2024). Water and sanitation for urban health: A gender perspective on impacts and coping strategies in Mangaung Metropolitan Municipality. In A. R. Matamanda & V. Nel (Eds.), *Sustainable development goals and urban health: Strides, challenges and way forward for poor neighborhoods* (pp. 49–67). Springer.

Musemwa, M. (2021). Urban struggles over water scarcity in Harare. *Daedalus, 150*(4), 27–47. https://doi.org/10.1162/daed_a_01871

Mutandwa, H., & Vyas-Doorgapersad, S. (2023). The state of water infrastructure development in Zimbabwe. *The African Journal of Governance and Development, 12*(2), 40–57. https://hdl.handle.net/10520/ejc-ajgd_v12_n2_a4

Rammile, S., Bhanye, J. I., Lehobo, T., & Hugo, M. (2024). Building resilient urban infrastructure and innovation for urban public health: A case study of Bloemfontein, South Africa. In A. R. Matamanda & V. Nel (Eds.), *Sustainable development goals and urban health: Strides, challenges and way forward for poor neighborhoods* (pp. 69–96). Springer.

Samhokore, A. (2022). Coping strategies and health implications of the urban poor in the face of limited water and sanitation services: The case of Caledonia in Harare Zimbabwe. (Doctoral dissertation, University of South Africa)

Taonameso, S., Mudau, L. S., Traoré, A. N., & Potgieter, N. (2022). Urban water conundrums in Zimbabwe–the role of water policy and its implementation gaps. *Water Supply, 22*(3), 3526–3539. https://doi.org/10.2166/ws.2021.386

Toriro, P. (2024). People-centred initiatives as a response to infrastructure collapse in Harare. In A. R. Matamanda & I. Chirisa (Eds.), *Urban infrastructure in Zimbabwe: Departures, divergences and convergences* (pp. 181–199). Springer.

UNICEF (United Nations Children's Fund). (2023). Cholera outbreak in Eastern and Southern Africa. https://www.unicef.org/esa/reports/cholera-outbreak-eastern-and-southern-africa-2023. Accessed: 04 September 2023.

United Nations. (2022). *The United Nations world water development report 2022 Groundwater: Making the invisible visible*. UNESCO. https://www.unwater.org/publications/un-world-water-development-report-2022

Water Research Commission. (2020). *Sanitation upgrading programmes: Lessons from the bucket eradication programmes*. Pretoria. https://www.wrc.org.za/wp-content/uploads/mdocs/Sanitation%20Upgrading%20-Bucket_Eradication_lesson.pdf

World Health Organization. (2020). *World health statistics 2020: monitoring health for the SDGs, sustainable development goals*. World Health Organization. https://iris.who.int/bitstream/handle/10665/332070/9789240005105-eng.pdf

World Health Organization. (2023). *Considerations for implementing for adjusting public health and social measures in the context of COVID-19*. Strategic Health Operations. https://iris.who.int/bitstream/handle/10665/341811/WHO-2019-nCoV-Adjusting-PH-measures-2021.1-eng.pdf

Yang, H., Xu, S., Chitwood, D. E., & Wang, Y. (2020). Ceramic water filter for point-of-use water treatment in developing countries: Principles, challenges and opportunities. *Frontiers of Environmental Science & Engineering, 14*, 1–10.

Zerbo, A., Delgado, R. C., & González, P. A. (2021). Water sanitation and hygiene in sub-Saharan Africa: Coverage, risks of diarrheal diseases, and urbanization. *Journal of Biosafety and Biosecurity, 3*(1), 41–45. https://doi.org/10.1016/j.jobb.2021.03.004

Zimstat (Zimbabwe National Statistics Agency). Zimbabwe Statistics 2022. Harare

Chapter 10
Conclusions and Future Directions: Pathways for Resilient and Sustainable Water Governance in Zimbabwe

Abstract This final chapter consolidates the key insights and discussions on urban water governance in post-colonial Zimbabwe. It synthesizes the key themes discussed throughout the book, emphasizing the impacts of historical practices, various institutions, political dynamics, economic constraints, and technological advancements on water management. The chapter highlights the need for sustainable and resilient governance frameworks that are adaptable to both current challenges and future uncertainties, especially those posed by climate change and socioeconomic shifts. Reflecting on the complex impacts of inadequate water governance—from public health to environmental sustainability—the chapter proposes strategic recommendations to enhance the resilience, efficiency, and equity of water resource management in Zimbabwe. This synthesis not only addresses the immediate imperatives for improved water service delivery but also sets forth a vision for a water-secure future, advocating for cross-scalar governance and a unified approach that encompasses community involvement, technological innovation, and robust policy reforms.

Keywords Urban water governance · Sustainability · Resilience · Zimbabwe · Climate change adaptation · Technological innovation · Policy reform

10.1 Introduction

In this concluding chapter, we synthesize the insights from the comprehensive analysis of urban water governance in postcolonial Zimbabwe. The preceding chapters of the book discussed diverse factors—from historical legacies, institutional dynamics, politics, finance, climate change, and gender, to modern-day technological advancements that influence and shape the governance of urban water resources in Zimbabwe. These discussions not only illuminated the complexities of managing such a vital resource but also highlighted the need for resilient and sustainable strategies that can withstand the pressures of both present and future challenges.

A. R. Matamanda et al., *Urban Water Governance in Postcolonial Zimbabwe*, Local and Urban Governance, https://doi.org/10.1007/978-3-031-96973-7_10

Our exploration began with an understanding of the historical context, tracing how colonial and postcolonial policies have left enduring marks on the nation's water governance structures. We moved into the political instrumentalization of water, revealing how governance is often compromised by political agendas that skew resource distribution and impair service delivery. The economic constraints, characterized by hyperinflation and budget misallocations, have been shown to exacerbate infrastructural decay, leading to significant service delivery shortfalls. Furthermore, the socioeconomic impacts of water scarcity, particularly on marginalized communities, were examined, underscoring the cycles of poverty and disease perpetuated by inadequate governance.

Environmental challenges, especially those posed by climate change, demand adaptive governance frameworks capable of responding to shifting weather patterns and water availability. The potential of technological innovations was discussed as a means to enhance both the resilience and efficiency of water service delivery systems. Finally, we addressed the water–health nexus, emphasizing the critical role of water quality in public health and the necessity for integrated approaches that align water policy planning with health outcomes.

The logic for adopting sustainable and resilient water governance practices in Zimbabwe is clear. Such governance not only addresses immediate needs but also ensures the long-term viability of water resources, balancing human demands with environmental preservation. It involves enhancing the robustness of infrastructure, diversifying water resources, and developing adaptive policies to effectively manage external pressures.

As we conclude this book, we aim to solidify the discourse on how Zimbabwe can develop a robust, equitable, resilient and sustainable water governance framework. This chapter not only recapitulates the main themes discussed but also sets forth strategic recommendations and future directions, urging all stakeholders to embrace these challenges with innovation, inclusivity, and a forward-looking perspective. Through collective responsibility and informed governance, Zimbabwe can achieve a water-secure future, ensuring that this essential resource remains accessible, safe, and plentiful for all citizens now and in the generations to come.

10.2 Summary of Key Themes, Findings and Lessons

10.2.1 Recap of the Main Themes Discussed in the Book

Throughout this book, we have examined several challenges and prospects that define urban water governance in postcolonial Zimbabwe, by unveiling the layers of complexity that shape the nation's water management strategies against a backdrop of socioeconomic, political, and environmental challenges.

The following key themes emerged from our discussions:

1. **Historical Context and Legacy Issues**: This theme provided a foundational understanding of how past policies and infrastructural developments continue to have an impact on current water governance. We traced the evolution of water management systems from colonial times to the present, emphasizing the cumulative effect of historical decisions on today's governance frameworks.
2. **Political Instrumentalization**: The book critically analyzed how water governance is often entangled with political agendas, affecting equitable resource distribution and efficient service delivery. It discussed the ramifications of using water access as a political tool, influencing policy decisions and management practices that prioritize political gain over public good.
3. **Economic Constraints and Opportunities**: Highlighting the economic underpinnings of water governance, this theme explored how economic instability—exacerbated by hyperinflation and budget misallocations—affects water infrastructure investment and maintenance, leading to service delivery shortfalls.
4. **Socio-Economic Impacts**: The book discussed the direct consequences of inadequate water governance on public health, economic productivity, and social equity. Particular attention was given to how water scarcity disproportionately affects marginalized communities, perpetuating cycles of poverty and disease.
5. **Environmental Challenges and Climate Change**: The book addresses the pressing issues of environmental degradation and climate change. It highlighted the critical need for adaptive water governance frameworks that can respond to changing weather patterns, water availability, and quality.
6. **Technological Innovations and Infrastructure Development**: This theme provided a forward-looking analysis of the role of technology in transforming water governance. We explored innovative solutions such as smart metering, sustainable infrastructure, and community-based water management systems that can enhance both resilience and efficiency in water service delivery.
7. **Health and Water Quality**: Delving into the water–health nexus, the discussions emphasized the critical role of water quality in public health. We examined how waterborne diseases like cholera and typhoid are exacerbated by poor governance and inadequate infrastructure, highlighting the urgent need for integrated approaches that consider health outcomes in water policy planning.

10.2.2 Summary of Critical Insights from Each Chapter

The insights from each chapter show the interconnected challenges of water governance in Zimbabwe, showing how historical, institutional, economic, and political factors converge to shape outcomes. These findings challenge the conventional management strategies and suggest that an integrated, multisectoral approach is required—one that considers not just the technical aspects of water governance but also the sociopolitical and economic contexts. Table 10.1 gives a summary of critical insights from each chapter.

10.2.3 *Lessons for Urban Water Governance, Sustainability, and Resilience in the Postcolonial Zimbabwean Context*

Table 10.1 Summary of critical insights from each chapter

Chapter	Insight	Impact
1. Introduction to urban water governance in a postcolonial context	The historical legacies of colonialism have shaped the frameworks and challenges of modern water governance, embedding deep structural issues	These historical contexts complicate contemporary governance efforts, necessitating tailored approaches to address legacy issues
2. Conceptual underpinnings on urban water governance	Various theoretical frameworks highlight the complexity of urban water governance, emphasizing the interplay of socioeconomic, political, and environmental factors	A multidisciplinary approach is crucial for effectively addressing the multifaceted challenges in water governance
3. Urban legislation and policy framework guiding water governance in postcolonial Zimbabwe	Legislation and policies have evolved but often remain ineffective due to lack of enforcement and alignment with local needs	There is a clear need for policy reforms that are both contextually appropriate and robustly enforced to improve water governance
4. Institutionalism and urban water governance in Zimbabwe	Institutional frameworks governing water are often fragmented and inefficient, affected by bureaucracy and corruption	Strengthening institutions is critical for enhancing accountability and the effectiveness of water governance
5. Unearthing the political economy and ecology of urban water governance in a failed state	Economic decline and ecological degradation have a severe impact on water governance, exacerbating the challenges in a politically unstable environment	Addressing the political economy and ecological issues is essential for sustainable water management
6. Climate change adaptation in urban water governance in the Zimbabwean context	Climate change poses significant risks to water availability and quality, requiring adaptive management strategies	Integrating climate adaptation into water governance is imperative to secure water sustainability for the future
7. The nexus between gender and urban water supply: A feminist political perspective	Gender inequalities are deeply embedded in water governance, with women disproportionately affected by water scarcity	Gender-sensitive approaches in water policy can improve access and equity in urban water supply
8. Infrastructure financing and technological advancements for urban water supply	Financial constraints and outdated technology hinder the development of sustainable water infrastructure	Innovative financing and technology are crucial for upgrading water infrastructure to meet growing urban demands
9. The political Instrumentaliza-tion of urban water governance in Zimbabwe: A water–health nexus	Political interests often manipulate water governance, affecting public health and exacerbating water-related diseases	Decoupling political influence from water governance is essential for improving public health outcomes and water service delivery

The following is a synthesis of key lessons drawn from the examination of urban water governance within the postcolonial context of Zimbabwe. These lessons are crucial not only for understanding the current challenges but also for designing more resilient, sustainable, and equitable water governance frameworks in Zimbabwe. Here, we articulate key takeaways that can guide policymakers, practitioners, and community stakeholders.

1. **Integration of Traditional and Modern Water Management Practices**: One significant lesson is the value of integrating traditional knowledge systems with modern water management practices. Zimbabwe, with its rich history of indigenous water management techniques like rain water harvesting and nature-based solutions, provides a fertile ground for blending these methods with contemporary technologies. This integration can lead to more culturally appropriate and widely accepted practices that enhance community engagement and resilience in water governance.
2. **Strengthening Institutional Coordination and Capacity**: Effective urban water governance requires robust institutional frameworks that are well-coordinated and capable of adapting to new challenges. Our findings highlighted the need for strengthening the capacities of local institutions, improving their coordination, and ensuring that they are adequately resourced. This involves streamlining the roles of various governmental and non-governmental actors to reduce overlaps and gaps in service delivery and accountability.
3. **Enhancing Public Participation and Transparency**: Public trust and effective governance are inextricably linked to the levels of stakeholder engagement and transparency. Lessons from Zimbabwe show that involving communities in the planning, decision-making, and implementation phases of water management projects lead to better outcomes. Transparency in administrative and financial operations is also crucial to combatting corruption and improving service delivery.
4. **Adopting Flexible Adaptive Management Approaches**: Climate variability and socioeconomic changes demand flexible and adaptive management approaches. Urban water governance systems should be designed to be dynamic, learning from past experiences and adjusting to future conditions. This adaptability is essential for managing uncertainties and shocks, such as droughts or economic crises that have a significant impact on water resources.
5. **Focusing on Infrastructure Investment and Maintenance**: The decay of water infrastructure is a critical issue that requires urgent attention. Sustainable financing models need to be developed to support the construction, renovation, and maintenance of water infrastructure. Investment should prioritize not only the physical aspects of infrastructure but also the technological upgrades necessary for efficient water management and distribution systems.
6. **Prioritizing Equity in Water Distribution**: Equity in water access remains a pressing challenge. Policies and practices should prioritize the provision of water services to underserved and marginalized communities. Ensuring equitable

water distribution involves addressing systemic inequalities and ensuring that all citizens have access to adequate, safe, and affordable water.

7. **Building Resilience Against Water-Related Risks**: Urban water governance must include strategies for mitigating and responding to water-related risks, including floods, waterborne diseases, and infrastructure failures. Developing robust emergency response strategies and regular risk assessments can help mitigate these risks and enhance the overall resilience of urban water systems.
8. **Leveraging International Partnerships and Learning**: International cooperation can provide critical support in terms of technical expertise, funding, and policy frameworks. Learning from global best practices on urban water governance and adapting them to the local context can accelerate improvements in water governance. Partnerships with international bodies dealing with water issues can also facilitate the exchange of knowledge and technologies that are beneficial for local development.

Through these lessons, this book aimed to contribute to the discourse on urban water management in similar contexts across the developing world. The challenges are undoubtedly complex, but the potential for transformative change through informed, inclusive, and innovative governance practices offers a hopeful vision for the future of urban water systems in Zimbabwe and beyond.

10.3 Importance of Sustainable and Resilient Water Governance in Zimbabwe

Sustainable and resilient water governance is not merely a policy preference but a critical necessity for Zimbabwe. The importance of this approach lies in its potential to address both immediate needs and long-term challenges, ensuring that all citizens have access to safe and reliable water services while safeguarding the environment for future generations.

Sustainability in water governance ensures that water use does not exceed its replenishment rates and that all aspects of water management—from extraction to treatment and distribution—are conducted in a manner that minimizes ecological impact (Pahl-Wostl et al. 2013). It calls for policies and practices that balance human needs with environmental preservation, promoting water conservation and the use of green technologies.

Resilience in water governance means the capacity of water management systems to recover from disruptions, adapt to changes, and continue to function effectively in the face of external pressures, including economic shocks, political changes, and climatic variations (Cosens & Williams, 2012). Building resilience involves enhancing the robustness of infrastructure, diversifying water sources, and developing adaptive policies that can respond to evolving circumstances.

The commitment to sustainable and resilient water governance is essential for Zimbabwe to overcome its current challenges and achieve its development goals. It provides a pathway to ensure that water—a fundamental human right and a critical resource for the country's development—remains accessible, safe, and plentiful for all Zimbabweans now and in the future.

10.3.1 Challenges to Resilient Water Governance in Zimbabwe

The journey towards resilient and sustainable water governance in Zimbabwe is fraught with numerous challenges that hinder the effective implementation of water management practices. These challenges stem primarily from systemic issues that impair the capacity of water systems to recover from disruptions and adapt to changing conditions (Mapedza et al., 2016). Persistent infrastructural inadequacies, compounded by inadequate maintenance and investment, have led to frequent system failures and inefficient water distribution (Chirisa et al., 2024). Additionally, management practices often lack the flexibility to adapt to the dynamic and evolving needs of urban populations, further complicating the provision of consistent and reliable water services.

Vulnerabilities in Zimbabwe's water governance are exacerbated by several key factors. Political instability remains a significant concern, as frequent changes in leadership and policy priorities disrupt consistent water management strategies and often lead to the misallocation of resources. Economic constraints also play a critical role, with limited funding available for the necessary upgrades and expansions of water infrastructure. This economic shortfall is often a result of broader macroeconomic challenges facing the country, including hyperinflation and fiscal deficits, which divert government focus away from essential service delivery (Bhanye et al., 2023).

Moreover, the decay of existing water infrastructure is a critical issue that contributes to the vulnerability of the water governance framework. Much of the infrastructure is outdated and has surpassed its useful life, leading to increased instances of breakdowns and service interruptions. This decay is seldom addressed with the urgency or scale required, due to the aforementioned economic constraints and political challenges, leading to a cycle of degradation and inefficiency that is difficult to break.

Together, these factors create a complex environment of challenges that impede the development of resilient and sustainable water governance systems in Zimbabwe. Addressing these issues requires a concerted effort from all stakeholders, including government, civil society, and international partners, to foster a governance environment that can withstand and adapt to the myriad challenges it faces.

10.3.2 Application of Principles of Sustainable Water Governance

Sustainability in urban water management means the ability to provide reliable and adequate water services in a manner that does not compromise the ability of future generations to meet their own needs (Pahl-Wostl et al., 2013). This concept encompasses various dimensions, including environmental, social, and economic sustainability. In the context of Zimbabwe, where urban water systems face numerous challenges, sustainability should involve developing practices that ensure long-term water security, promote conservation, and minimize environmental degradation. The sustainable management of urban water systems should also involve integrating water supply with urban development planning to ensure that, as cities expand, water services can adapt and continue to meet the needs of growing populations without exacerbating the existing pressures on water resources. Table 10.2 presents the core principles of sustainable urban water governance.

Table 10.2 Core principles of sustainable water governance in urban water systems

Principle	Description
Equity	Ensures that all individuals have fair access to water services, regardless of their socioeconomic status. This principle addresses disparities in water availability and quality across different urban areas and among various socioeconomic groups. In Zimbabwe, achieving equity should involve rectifying imbalances where marginalized communities often suffer from inadequate access to clean water and sanitation facilities.
Efficiency	Pertains to the optimal use of available water resources and the minimization of water waste. Efficient water governance involves the adoption of technologies and practices that reduce water losses, such as leak detection, repair programs, and water-efficient appliances. It also includes the efficient allocation of financial resources to areas where they can achieve the greatest impact in terms of service delivery improvement and infrastructure resilience.
Sustainability	Emphasizes the importance of managing water resources in a way that maintains their availability for future generations. It involves protecting and conserving water sources through pollution control, sustainable withdrawal practices, and the implementation of green infrastructure solutions like rainwater harvesting and wastewater recycling.
Adaptability	Refers to the capacity of water systems to adjust to changing conditions, including climatic variations, population growth, and economic shifts. This principle is particularly relevant in the face of climate change, which poses additional risks such as increased frequency of droughts and floods. Water governance structures must be flexible and responsive, able to modify policies and practices as circumstances evolve.

10.4 Strategic Recommendations for Improvement

10.4.1 Recommendations to Address the Immediate and Long-Term Challenges of Water Access and Quality in Urban Zimbabwe

This section presents strategic approaches to address the immediate and long-term challenges of water access and quality in Zimbabwe's urban areas (Table 10.3). The strategies are categorized into short-term actions, which focus on immediate relief and efficiency improvements, and long-term strategies, which aim at sustainable

Table 10.3 Strategic approaches to enhancing water access and quality in Zimbabwe

Strategy type	Strategy	Description
Short term	Emergency water supply initiatives	Deploy mobile water purification units and water tankers to areas experiencing acute water shortages, providing immediate relief to populations suffering from water scarcity
	Leak detection and repair programs	Rapidly implement programs to identify and repair leaks in the urban water distribution network to reduce non-revenue water and improve the efficiency of water delivery systems
	Public health campaigns	Enhance public health through campaigns promoting water purification methods at the household level, such as boiling, using chlorine tablets, or solar disinfection techniques to improve water safety temporarily
	Strengthening water rationing schemes	Implement fair and transparent water rationing schedules that are well communicated to the public to manage limited resources more effectively during times of scarcity
Long term	Infrastructure development	Invest in upgrading and expanding water treatment facilities and distribution networks, including the construction of new reservoirs, rehabilitation of old pipes, and installation of modern water purification technologies to increase capacity and efficiency
	Policy reforms	Reform policies to ensure water governance is inclusive, equitable, and efficient. This may include revising water tariffs to ensure they are fair but also fund maintenance and expansion, establishing clear regulations for water quality, and implementing policies that encourage water conservation and reuse
	Integrated water resources management IWRM)	Adopt IWRM approaches that consider the entire water cycle, from source to sea, and integrate water management practices across sectors and stakeholders to enhance the sustainability of water resources
	Stakeholder engagement	Develop mechanisms for robust community and stakeholder engagement in water management decisions. This should include regular consultations with urban residents, especially marginalized communities, to ensure their needs and voices are considered in water governance

infrastructure development, comprehensive policy reforms, and enhanced stakeholder engagement. These efforts are critical to ensure that all citizens have access to safe and reliable water services, while also laying the groundwork for a resilient and sustainable water management system in the future.

10.4.2 Recommendations for Enhancing Resilience Against Future Water-Related Crises

Policymakers and practitioners must focus on building adaptive capacity and preparedness to enhance resilience against future water-related crises:

1. **Climate–Resilient Infrastructure:** Upgrade infrastructure to withstand extreme weather events such as floods and droughts. This may include elevating critical infrastructure, using flood-resistant materials, and creating green spaces that can absorb excess rainwater.
2. **Early Warning Systems and Emergency Preparedness:** Implement advanced early warning systems for droughts and floods and establish clear protocols for emergency responses to water crises, ensuring that responses are swift and effective.
3. **Diversifying Water Sources:** Explore alternative water sources such as rainwater harvesting, reclaimed wastewater for non-potable uses, and desalination where feasible. This diversification can provide a buffer capacity in the water system, reducing reliance on any single water source.
4. **Capacity Building and Education:** Invest in education and training for local authorities and the general public on sustainable water use, crisis management, and adaptation strategies to build a more informed and resilient community.

10.5 Role of Technology and Innovation

10.5.1 Innovative Solutions for Water Purification, Waste Management, and Distribution Systems

Technology and innovation play critical roles in transforming urban water management systems, especially in the areas of water purification, waste management, and distribution. Innovative water purification technologies such as reverse osmosis, ultraviolet disinfection, and advanced filtration systems can significantly improve the quality of water provided to urban populations (Guo et al., 2022). These technologies are capable of removing contaminants, pathogens, and other pollutants more efficiently, ensuring that the water meets safety standards suitable for human consumption.

In waste management, innovations like anaerobic digestion and membrane bioreactors enhance the treatment of wastewater, allowing for the safe reuse of water in agricultural and industrial applications. This not only conserves freshwater resources but also reduces the environmental impact of untreated wastewater. Moreover, modernizing the water distribution systems through the installation of smart piping solutions can reduce leaks and bursts. These smart pipes are embedded with sensors that detect changes in pressure and flow rates, enabling real-time monitoring and rapid response to potential leaks, significantly reducing water loss and ensuring a more consistent water supply.

10.5.2 Smart Technologies and Data Management Potential in Enhancing Water Governance

Smart technologies and data management are revolutionizing water governance by providing detailed insights into water usage, quality, and distribution efficiency (Nguyen et al., 2018). Smart water meters, for example, are instrumental in promoting water conservation among consumers. These meters provide real-time data on water consumption, allowing users and utilities to monitor usage patterns and identify areas for reduction. Utilities can use this data to implement dynamic pricing models that encourage water-saving behavior, especially during peak demand or drought conditions.

Furthermore, GIS and remote sensing technologies offer powerful tools for managing water resources more effectively (Fernández Moniz et al., 2020). GIS platforms can integrate various data types, including spatial data on water infrastructure, consumer demographics, and environmental conditions, facilitating better planning and management decisions. This comprehensive data visualization supports the optimization of water distribution routes, the identification of critical areas for infrastructure investment, and the assessment of risks related to water scarcity and contamination.

Data management also plays a crucial role in enhancing the operational efficiency of water systems. Through utilizing big data analytics, water utilities can predict demand trends, optimize treatment processes, and enhance maintenance schedules. Predictive analytics can forecast potential system failures or disruptions, allowing for preemptive maintenance and reducing downtime and repair costs.

The integration of IoT technologies in water management systems further enhances monitoring and control capabilities. IoT devices can continuously collect data from various points in the water supply network, providing insights into the system's performance and environmental conditions. This enables water managers to make informed decisions fast, enhancing responsiveness to changes in water quality or supply needs.

10.6 Community Engagement and Empowerment

Community engagement is fundamental in shaping water governance systems that are not only inclusive but also effective and responsive to local needs (Mgwele et al., 2024). Involving communities in the planning and decision-making processes ensures that the solutions developed are tailored to address specific challenges faced by those directly affected by water issues. This participatory approach fosters a sense of ownership and responsibility among community members, enhancing the sustainability of water projects. Moreover, when communities are actively involved, they can provide valuable insights and local knowledge that can lead to more innovative and practical solutions.

Empowering communities to manage their water resources effectively requires a combination of education, resource allocation, and supportive policy frameworks. Education and training programs are essential in building local capacity to manage water systems. These programs should cover topics such as sustainable water use, basic water treatment techniques, system maintenance, and governance principles. Providing communities with the necessary skills and knowledge empowers them to take active roles in the management of their water resources.

Moreover, allocating resources such as tools, materials, and funding to community water projects can catalyze their involvement and commitment to maintaining water infrastructure. This could include support for constructing community wells, installing purification systems, or developing rainwater collection systems, tailored to the community's specific needs and environmental conditions.

Finally, supportive policies that encourage community governance of water resources can significantly empower local groups. Policies that facilitate the establishment of community water management committees or local governance structures enable residents to make decisions about water distribution, pricing, and maintenance. These committees can also serve as platforms for advocacy, where communities can communicate their needs and challenges to higher levels of government, ensuring that their voices are heard and acted upon.

10.7 Partnerships and Collaborations

Partnerships and collaborations play a critical role in enhancing the effectiveness of water governance frameworks, both at national and international levels. These partnerships bring together diverse expertise, resources, and perspectives that are crucial for addressing the complex challenges of water management in an integrated and sustainable manner (Margerum & Robinson, 2015). National partnerships frequently involve various government departments working together to ensure coherence in policies and practices related to water management. Internationally, partnerships can provide access to technical expertise, financial resources, and global best practices that can be adapted to local contexts.

Collaborative efforts between the government, private sector, NGOs, and international organizations have the potential for a significant impact on water governance. Governments can establish the regulatory frameworks and policies that facilitate effective water management while providing oversight and coordination. The private sector can offer innovation, efficiency, and capital investment, particularly in the areas of water technology, infrastructure development, and management services. NGOs often play a crucial role in advocacy, raising awareness, community engagement, and implementing grassroots projects that address specific local water issues. International organizations and bodies can provide funding, technical expertise and a platform for sharing knowledge and experiences across borders.

For example, a partnership between a national government agency, a global environmental NGO, and a private technology firm could develop a project to implement advanced water purification technologies in underserved urban areas. The government could facilitate the regulatory approvals and support needed, the NGO could manage the community engagement and education components, and the private firm could provide the technology and expertise for installation and maintenance.

10.8 Policy and Regulatory Frameworks Impact on Water Governance

In Zimbabwe, the framework governing water resources is constituted under various policies and acts, such as the Water Act and the Environmental Management Act. These legislations aim to manage and preserve water resources, but they face significant challenges due to implementation gaps, lack of coordination between governing bodies, and insufficient enforcement mechanisms. The current policies have been somewhat effective in setting the groundwork for water governance, yet they often fall short in addressing the needs of a rapidly changing environment and diverse urban demands. The impact of these policies is mixed; while they have established basic water rights and management structures, they struggle to cope with issues such as pollution control, water distribution inequalities, and infrastructure decay.

To enhance sustainability and resilience in water governance, Zimbabwe requires a robust overhaul of its policy framework. New regulations should focus on integrating climate resilience into all aspects of water management, ensuring that water governance frameworks can adapt to and withstand the impacts of climate variability and change. For instance, incorporating mandatory water-saving measures and sustainable extraction practices could help mitigate the effects of droughts and water scarcity.

Additionally, there is a critical need for policies that promote the equitable distribution of water resources. This could be facilitated by revising tariff structures to ensure that water pricing is fair, yet encourages conservation. Regulations should also enforce stricter penalties for pollution to protect water quality for all users.

New policies must prioritize the expansion and maintenance of water infrastructure to reduce losses and improve efficiency in the water supply network.

Effective policy frameworks are crucial for driving innovation and investment in the water sector. By setting clear guidelines and providing incentives for research and development, policies can encourage the adoption of advanced water technologies such as smart metering systems, wastewater recycling processes, and efficient irrigation techniques. For example, the government could introduce tax breaks or subsidies for companies that develop and implement water-saving technologies or for property developers who integrate sustainable water management systems into their projects. Moreover, policies that facilitate public–private partnerships are essential to mobilize the capital necessary for large-scale water infrastructure projects. Such collaborative frameworks can attract domestic and international investors by providing a stable and predictable environment for their investments. Regulatory support is vital for fostering an environment where innovative solutions can be tested and scaled, ultimately leading to significant improvements in water management practices.

10.9 Monitoring, Evaluation, and Learning

10.9.1 Systems and Processes for Ongoing Monitoring and Evaluation

Effective water governance requires robust systems for ongoing monitoring and evaluation (M&E) to ensure that water resources are managed efficiently and sustainably. Establishing comprehensive M&E systems involves the regular collection of data on water quality, usage patterns, and infrastructure condition (Larson et al., 2016). This data should be systematically analyzed to assess the performance of water management strategies and the condition of water resources. Tools such as GIS and remote sensing can provide vital spatial data that enhance the monitoring of water bodies and infrastructure, facilitating timely interventions. In Zimbabwe, integrating advanced data collection technologies with traditional monitoring methods will enable more precise and comprehensive management of water resources. Regular audits and performance evaluations of water projects should be mandated to maintain high standards of accountability and effectiveness in water service delivery.

10.9.2 Importance of Adaptive Management and Learning in Water Governance

Adaptive management is a critical component of effective water governance, allowing policymakers and managers to modify strategies in response to new information and changing conditions. This approach is particularly important in the context of climate change and rapid urbanization, which introduce new uncertainties and risks in water management. For adaptive management to be successful, it must be underpinned by a strong learning culture within water governance institutions. This involves continuously testing assumptions, learning from past experiences, and incorporating lessons learned into future planning and policymaking. Training and capacity-building programs are essential to equip water management professionals with the skills needed to adapt to new challenges and technologies.

10.9.3 Using Feedback Mechanisms to Improve Water Governance Strategies

Feedback mechanisms are vital for connecting the various stakeholders involved in water governance, including government agencies, private companies, NGOs, and community members. Effective feedback systems ensure that all stakeholders can express concerns, provide input, and share experiences and knowledge. By establishing formal channels such as stakeholder forums, user committees, and online platforms in Zimbabwe, it can facilitate the flow of feedback from all sectors of society. These platforms should encourage transparent communication and include mechanisms for stakeholders to contribute to decision-making processes. Feedback from these interactions should be systematically integrated into the monitoring and evaluation systems to refine and improve water governance strategies. For example, community feedback can reveal issues with water access that might not be visible through conventional monitoring. Such insights can drive quick policy adjustments to address disparities in water distribution or problems with infrastructure. Similarly, feedback from private sector participants can help identify innovative solutions and investment opportunities that could be leveraged to enhance water system efficiencies.

10.10 Future Directions and Research Needs

10.10.1 Emerging Trends and Future Research Areas in Water Governance

Water governance is an evolving field influenced by technological advances, climate change, and socioeconomic transformations. Future research in this area must address the integration of new technologies, such as AI and IoT, into water management systems to enhance efficiency and responsiveness. These technologies can facilitate real-time data collection and analysis, offering unprecedented insights into water usage, leakage detection, and pollution control. Research should also explore the socioeconomic impact of water governance policies, particularly how they affect marginalized and vulnerable populations. Another critical area of research is the impact of climate change on water resources. As Zimbabwe faces increased variability in rainfall and more frequent droughts, understanding these patterns and developing strategies to manage water scarcity is crucial. Studies focusing on sustainable water sourcing, such as rainwater harvesting and the reuse of treated wastewater, will be increasingly important.

10.10.2 Anticipated Challenges and Opportunities in the Coming Decades

The next few decades will bring significant challenges for water governance due to population growth, urbanization, and the intensification of agricultural activities. These pressures will strain existing water resources and infrastructure, necessitating innovative approaches to water management and conservation. Opportunities may arise from advancements in water purification and desalination technologies, which could make previously unusable water sources viable options. Policy development must also consider the transboundary nature of water resources. As Zimbabwe shares water bodies with neighboring countries, regional cooperation and agreements will be vital to ensure the equitable and sustainable use of these shared resources. Additionally, the potential impacts of global economic shifts on funding for water infrastructure projects present both a challenge and an opportunity to advocate for increased investment in sustainable water systems.

10.10.3 Call for Ongoing Research and Dialogue to Adapt and Respond to Future Changes

To effectively navigate the complexities of future water governance, there must be a concerted effort to foster ongoing research and dialogue among all stakeholders, including policymakers, researchers, community leaders, and international partners. This dialogue should aim to share knowledge, coordinate actions, and build consensus on best practices and policies. Research institutions and universities in Zimbabwe should develop specialized programs and partnerships that focus on water governance. These programs can serve as incubators for new ideas and technologies, facilitating the translation of research findings into practical applications and policy recommendations. Moreover, international collaboration can provide access to new insights, funding, and technologies that may not be readily available domestically. Engaging with global water governance networks can help Zimbabwe stay abreast of best practices and emerging trends, ensuring that its water governance strategies are informed by the latest research and successful case studies from around the world.

10.11 Conclusion

This book discussed and outlined a comprehensive pathway towards achieving resilient and sustainable water governance in Zimbabwe, emphasizing the integration of modern technologies, community involvement, and robust policy frameworks. We explored the complex relationship between political dynamics, infrastructural needs, and the socioeconomic factors that influence water governance in postcolonial Zimbabwe. The discussions highlighted the urgent need for reforms that cater to both immediate and long-term water governance challenges, ensuring that all citizens have access to safe and adequate water resources.

The pathway to resilience and sustainability in water governance requires a multifaceted approach. It necessitates significant investment in infrastructure, the adoption of cutting-edge technologies for water management, and the strengthening of regulatory frameworks that ensure equity and efficiency. Furthermore, the active participation of communities in the planning and decision-making processes has been shown to enhance the effectiveness and sustainability of water governance systems, fostering greater accountability and responsiveness from governmental bodies.

The management of water resources is not just a governmental duty but a collective responsibility that involves every segment of society. This collective responsibility means that individuals, communities, businesses, and public officials must work collaboratively to effectively steward the country's water resources. It requires a shared commitment to sustainable practices, conservation efforts, and equitable policies that address the needs of all, especially the most vulnerable populations.

Incorporating the knowledge and experiences of local communities into water governance can bridge the gap between policy and practice, ensuring that solutions are culturally appropriate and technically feasible. Moreover, businesses and industries must adopt more sustainable practices that minimize water wastage and pollution, contributing positively to the overall health of the nation's water systems.

As we look towards the future, the role of governance in securing a sustainable future for Zimbabwe's water cannot be overstated. Effective water governance acts as the backbone of any strategy aimed at sustainability and resilience. It is through wise governance that resources can be allocated efficiently, technologies implemented effectively, and communities empowered to act as stewards of their environment.

Let us be inspired by the possibility of a future where water—our most precious resource—is managed with the care, respect, and foresight it deserves. Imagine a Zimbabwe where every citizen can access clean water, where every child can grow up free from the threat of waterborne diseases, and where every community can thrive because its water needs are met sustainably. This vision is attainable with the continued commitment to the principles of equity, sustainability, adaptability, and efficiency in our approach to water governance.

Together, we have the power to forge a water-secure future for Zimbabwe, ensuring that our cherished natural resources are preserved for generations to come. Let this book serve as a call to action, motivating all stakeholders to embrace the challenge of governance with the knowledge that their efforts will culminate in a legacy of abundance and health. Let us move forward with the collective resolve to ensure that water—as a fundamental right and a critical asset—is governed with the wisdom and dedication that its importance demands.

References

Bhanye, J. I., Mangara, F., Matamanda, A. R., & Kachena, L. (2023). *COVID-19 lockdowns and the urban poor in Harare, Zimbabwe*. Springer. https://doi.org/10.1007/978-3-031-41669-9

Chirisa, I., Ndemo, N., Toriro, P., & Chigudu, A. (2024). Urban expansion and sustainable water argumentation for Africa's cities. In R. Brears (Ed.), *The Palgrave encyclopedia of sustainable resources and ecosystem resilience* (pp. 1–16). Springer.

Cosens, B. A., & Williams, M. K. (2012). Resilience and water governance: Adaptive governance in the Columbia River basin. *Ecol Soc, 17*(4) https://www.jstor.org/stable/26269198

Fernández Moniz, P., Almeida, J. S., Pino, A. T., & Suárez Rivero, J. P. (2020). A GIS-based solution for urban water management. *Water Int, 45*(6), 660–677. https://doi.org/10.1080/02508060.2020.1765130

Guo, K., Wu, Z., Chen, C., & Fang, J. (2022). UV/chlorine process: An efficient advanced oxidation process with multiple radicals and functions in water treatment. *Acc Chem Res, 55*(3), 286–297. https://doi.org/10.1021/acs.accounts.1c00269

Larson, S., Kirono, D. G. C., Tjandraatmadja, G., & Barkey, R. (2016). Monitoring and evaluation approaches in water resources project design: Experiences from an urban water system climate change adaptation project in Indonesia. *Water Policy, 18*(3), 708–726. https://doi.org/10.2166/wp.2015.144

Mapedza, E., Manzungu, E., Rosen, T., Ncube, P., & Van Koppen, B. (2016). Decentralised water governance in Zimbabwe: Disorder within order. *Water Resources and Rural Development, 8*, 1–11. https://doi.org/10.1016/j.wrr.2016.05.001

Margerum, R. D., & Robinson, C. J. (2015). Collaborative partnerships and the challenges for sustainable water management. *Current Opinion in Environmental Sustainability, 12*, 53–58. https://doi.org/10.1016/j.cosust.2014.09.003

Mgwele, A., Bhanye, J. I., & Mocwagae, K. S. (2024). Water and sanitation for urban health: A gender perspective on impacts and coping strategies in Mangaung metropolitan municipality. In A. R. Matamanda & V. Nel (Eds.), *Sustainable development goals and urban health: strides, challenges and way forward for poor neighborhoods* (pp. 49–67). Springer.

Nguyen, K. A., Stewart, R. A., Zhang, H., Sahin, O., & Siriwardene, N. (2018). Re-engineering traditional urban water management practices with smart metering and informatics. *Environmental Modelling & Software, 101*, 256–267. https://doi.org/10.1016/j.envsoft.2017.12.015

Pahl-Wostl, C., Arthington, A., Bogardi, J., Bunn, S. E., Hoff, H., Lebel, L., Nikitina, E., Palmer, M., Poff, L. N., Richards, K., Schlüter, M., Schulze, R., St-Hilaire, A., Tharme, R., Tockner, K., & Tsegai, D. (2013). Environmental flows and water governance: Managing sustainable water uses. *Current Opinion in Environmental Sustainability, 5*(3–4), 341–351. https://doi.org/10.1016/j.cosust.2013.06.009

The manufacturer's authorised representative in the EU is Springer Nature Customer Service Centre GmbH, Europaplatz 3, 69115 Heidelberg, Germany. If you have any concerns regarding our products, please contact ProductSafety@springernature.com

Printed and bound by CPI Group (UK) Ltd, Croydon, CR0 4YY
01/10/2026
02225961-0005